MULTIPHASE PRODUCTION

IFP PUBLICATIONS

▶ J. FALCIMAIGNE

▶ S. DECARRE

MULTIPHASE PRODUCTION

Pipeline Transport, Pumping and Metering

2008

Editions TECHNIP 25 rue Ginoux, 75015 PARIS, FRANCE

FROM THE SAME PUBLISHER

- Basin Analysis and Modeling of the Burial, Thermal and Maturation Histories in Sedimentary Basins
  M. MAKHOUS, Y. GALUSKIN
- Sedimentary Geology
  Sedimentary basins, depositional environments, petroleum formation
  B. BIJU-DUVAL
- A Geoscientist's Guide to Petrophysics
  B. ZINSZNER, F.M. PELLERIN
- Geomechanics in Reservoir Simulation
  P. LONGUEMARE
- Oil and Gas Exploration and Production
  Reserves, costs, contracts
  CENTRE OF ECONOMICS AND ADMINISTRATION (IFP-SCHOOL)
- Integrated Reservoir Studies
  L. COSENTINO
- Geophysics for Sedimentary Basins
  G. HENRY
- Basics of Reservoir Engineering
  R. COSSÉ
- Well Seismic Surveying
  J.L. MARI, F. COPPENS
- Geophysics of Reservoir and Civil Engineering
  J.L. MARI, G. ARENS, D. CHAPELLIER, P. GAUDIANI

Printed in France

ISBN 978-2-7108-0913-5

Foreword

The methods and systems for producing hydrocarbon fluids have changed significantly over the last two decades. This change was driven by a number of factors:

- less frequent discoveries of large fields in mature basins such as in the North Sea or in the Gulf of Mexico, while many small fields were available for development if cost-effective methods could be found;
- increased water depth in offshore developments (especially in Gulf of Mexico, Brazil, Gulf of Guinea) requiring new technical approaches;
- a general need to reduce development costs in periods of low oil prices, as during the second half of the 80's.

These factors led the oil companies to reconsider conventional field development schemes, especially offshore. Conventionally, the gas and oil are separated and stabilized to commercial specifications before transportation by pipeline or tanker.

To reduce the complexity of full development and limit the facilities to small unattended platforms or subsea stations, production is now exported in multiphase flow conditions towards a central processing facility, which either exists or is created to serve several new fields. This scheme significantly improves the economics of marginal fields.

Tie-back distances between the wells and the remote-treatment installation were initially limited to no more than 10 to 15 kilometres. The tie-back distance was gradually extended and this trend is still continuing. For deepwater gas fields, for instance, the produced gas which includes some liquids (oil condensates and water produced during transportation under the combined effects of cold temperatures and head losses in pipelines) can be directly exported to the shore for treatment in facilities at distances of up to140 km.

Fluids which flow from the reservoir are composed of a mixture of oil, gas, water, often with corrosive components, sometimes laden with solid particles (sand) or scales. They are a potential source of many flow instabilities or problems such as liquid and gas slugs, severe slugging, flow line plugging by the formation of gas hydrates or deposition of wax, scales, asphaltenes, accumulation of sand. Over long distances, the natural pressure of the reservoir is not sufficient to balance the pressure drop in the pipeline, and the multiphase mixture needs to be boosted. An adequate instrumentation system is also needed to monitor and control the performance of such a system, and particularly the means to meter the oil, gas and produced water flowrates in multiphase conditions.

To solve these problems and make the required technology available, research and development programmes were undertaken in several areas. The main technical areas are:

- to improve knowledge of multiphase flow behaviour in pipelines and the accuracy of flow predictions by numerical calculations,
- to solve the physico-chemistry problems, particularly the risk of hydrate formation,
- to develop specific multiphase pump systems and multiphase flow meters.

The main purpose of the book is to present the fundamentals of multiphase production with regard to multiphase pipeline simulation, multiphase pumping and multiphase metering. Problems related to the physico-chemistry of produced fluids, hydrate and wax control, sand transportation etc. are not included in this book, as by themselves they would require a complete volume.

The essential purpose of this book is to provide engineers involved in field development and petroleum engineering students with information on approaches and technologies which can be used today. Special attention was been given to providing a detailed bibliography which could be used to examine a specific subject more closely.

This led the authors to devote specific chapters to pipeline flow simulations, multiphase pumping and multiphase metering. In the last chapter, a number of R&D works dealing with various areas are briefly presented and discussed.

- Pipeline flow simulations:

Today, the production of gas or oil leads to the study of a complex mixture composed of gas, liquid and solid phases. To design a field development, hydrodynamic and thermal studies are necessary. The chapter presents a description of the multiphase flow patterns that may appear in a pipe and explains how they are detected by the models. It also gives a good introduction to modelling of hydrodynamic effects, thermodynamics and thermal exchanges.

- Multiphase pumping:

Multiphase pumps are now deployed all round the world. More than 500 units are installed in fields, whether onshore, offshore or subsea. Their size has also considerably increased since the first prototypes were tested. The chapter presents the various multiphase pump concepts and explains their operating principles and how they can be introduced into a production system. A few application cases are described in the last section of the chapter.

- Multiphase flow meters:

Although the accuracy of multiphase flow meters may appear relatively poor in relation to single-phase flow metres, it is now considered to be acceptable for reservoir management and production purposes, and as accurate as well testing with conventional test separators. There are also a number of advantages such as being able to perform real-time monitoring of the field. They are therefore increasingly used in place of conventional test separators, not only offshore but also onshore. The chapter provides the basic principles of operation which relies on a combination of elementary measurements. It also describes the usual way of reporting operational performance and meter accuracy. An overview of a few field experiments is also given in this chapter.

- New challenges:

The successful experimentation with long-distance multiphase transportation, and multiphase pumping and metering has paved the way for the development of several other types of subsea processing: transportation of hydrates in slurry, subsea separation with re-injection of produced water, direct injection of raw seawater into the reservoir, subsea gas compression, energy or recovery with a multiphase turbine, etc. Several R&D programs are underway to extend the present state-of-the-art towards complex handling and processing of multiphase fluids. The requirements of oil and gas production in very deep water is a strong driver for these developments.

Acknowledgements

The book was prepared, written and reviewed with the help of several people, who provided comments, advice, or scientific and technical material. The authors wish to acknowledge the contributions of:

Dr. Eric Heintzé
Dr. Jacques Jarrin
Dr. Philippe Ungerer
Dr. Michel Constant
Dr. Yves Charron

The authors are particularly indebted to three of their colleagues who contributed to the writing of several parts of the book:

Dr. Lachet Veronique	paragraph 1.3.5 (pp. 22-34)
Dr. Duret Emmanuel	paragraph 1.3.6 (pp. 35-41)
Dr. Vilagines Regis	paragraphs 2.6 and 3.7 (pp. 81-90 & 124-131)

The authors reserve special thanks for Mr. Oddo Jean-Michel (Sulzer Pumps) for his kind help, and they express their regret and apologies for not including the proposed subject within the scope of the book.

The authors also wish to thank all those who contributed indirectly by developing the scientific subjects and the technology described in the book and who are acknowledged in the reference list.

Symbols

Roman symbols

A	pipe cross sectional area	m^2
a	intermolecular attraction parameter of the eos	$J\ m^3\ mol^{-1}$
b	covolume parameter of the eos	$m^3\ mol^{-1}$
Cp	heat capacity at constant pressure	$J\ K^{-1}\ mol^{-1}$
D	internal pipe diameter	m
Dh_k	hydraulic diameter for phase k	m
d	bubble diameter	m
e_k	internal energy per unit of mass of phase k	J/kg
f_k	internal fugacity	Pa
g	acceleration due to gravity	m/s^2
G	excess Gibbs energy	J
H	excess enthalpy	J
h_{fluid}	heat-transfer coefficient	$Wm^{-2}K^{-1}$
j_k	superficial velocity of phase k	m/s
k	binary interaction parameter	–
K	equilibrium coefficient	–
k_w	pipe roughness	m
MM	molar mass	$kg\ mol^{-1}$
$\mathbf{n}_{kj}$	normal unit vector	–
$\mathbf{n_z}$	unit vector on the axis of the pipe	–
n_i^k	number of moles of component i in phase k	mol
P	absolute pressure	Pa
P_{wk}	wall perimeter in contact with phase k	m
P_{ik}	interfacial perimeter between phases i and k	m
Q_{in}^W	heat flux per unit length	Wm^{-1}
R	gas constant	$J\ K^{-1}\ mol^{-1}$

T	absolute temperature	K
$\mathbf{u}_k$	local velocity of phase *k*	m/s
$\mathbf{U}_k$	mean value of the velocity of phase *k*	m/s
U_{Lfilm}	liquid velocity in the film region (slug flow)	m/s
U_{M}	mixture velocity	m/s
U_{T}	transitional velocity of the elongated bubble nose	m/s
U_∞	terminal bubble velocity	m/s
$\tilde{\mathbf{u}}_k$		m/s
v	molar volume	$m^3\ mol^{-1}$
V	total Volume	m^3
V_k	volume of phase *k*	m^3
V_{FR}	net gas entrainment rate at the slug front	m^3/s
$\dot{V}_{FR}$	net gas entrainment flux at the slug front	m/s
W_{i}	interfacial width	m
x	molar fraction in the liquid phase	–
y	molar fraction in the vapour phase	–
Z	compressibility factor ($= PV/nRT = Pv/RT$)	–

Greek symbols

α_k	phase fraction of phase *k* in Chapters 2 & 3	–
ε_k	phase fraction of phase *k*	–
$\varepsilon_{\mathrm{r}k}$	relative permittivity of phase *k*	–
$\varepsilon_{\mathrm{dis}}$	dissipation per unit mass	–
γ_i^k	activity coefficient of component i in the phase k	–
φ_i^k	fugacity coefficient of component i in the phase k	–
μ_k	dynamic viscosity of phase *k*	Pa.s
ν_k	kinematic viscosity of phase *k*, μ_k/ρ_k	m^2/s
θ	pipe inclination angle	rad
ρ_k	density of phase *k*	m^2/s^3
σ_{GL}	gas-liquid surface tension	N/m
$\tau_{\mathrm{w}k}$	phase *k* – wall shear stress	N/m^2
τ_{ik}	interfacial shear stress between phases *i* and *k*	N/m^2
ω	acentric factor	–

Subscripts

c	critical parameter
G	gas phase
i	component index in a mixture
L	liquid phase
M	mixture (sum of all the phases)
V	vapour phase
σ	saturation property

Abbrevations

eos	equation of state
BEP	Best efficiency point
BWR	Benedict-Webb-Rubin equation of state
CPA	Cubic Plus Association equation of state
ESP	Electric submersible pump
GVF	Gas volumetric fraction
GLR	Gas Liquid Ratio
LLE	Liquid- Liquid Equilibrium
MBWR	Modified Benedict-Webb-Rubin equation of state
MPFM	Multiphase Flowmeter
MPP	Multiphase Pump
NPSH	Net positive suction head
NRTL	Non Random Two Liquids GE model (Renon & Prausnitz, 1968)
PCP	Progressing Cavity Pump
PR	Peng-Robinson equation of state
PRH	Peng-Robinson equation of state with the Huron-Vidal mixing rule
SAFT	Self-Associating Fluid Theory equation of state
SMUBS	Shell multiphase booster station
SRK	Soave-Redlich-Kwong equation of state
SW	Soreide-Whitson equation of state
UNIQUAC	Universal Quasi Chemical model
VdW	Van der Waals equation of state

Table of Contents

Chapter 1
MULTIPHASE FLOW IN PIPELINES

Chapter 2
MULTIPHASE PUMPING

Chapter 3
MULTIPHASE METERING

Chapter 4

NEW CHALLENGES

Chapter 1

Multiphase Flow in Pipelines

1.1 INTRODUCTION

During the field design, hydrodynamic studies are performed to define the equipment between the reservoir and the topsides. The complex mixtures that are transported generally lead to critical flow problems, as is for example the case with hydrates, paraffin, heavy oils, etc. Starting from a production scenario and fixed input data, the determination of pipe diameters, the temperature evolution in shutdown, natural flow limits and restart problems are studied. The main goal of these studies is to optimise the pipe diameter in order to achieve minimum pressure drop keeping the fluid velocity below the erosion velocity. In the thermal studies, the temperature evolution is defined with respect to the insulation that has been chosen.

Depending on the nature of the field, pigging operations and production activation will have to be designed (gas lift, multiphase boosting, subsea separation).

To illustrate the importance of the design, two examples are given below:

1) A 3,000 m multiphase production riser from the reservoir (350 bar). Because of the pressure loss, the well head pressure will be around 20 bar.

The export line consists of a downward vertical riser (3,000 m), a horizontal part and an upward vertical riser (3,000 m) arriving at the process unit.

The actual tools predicting the pressure loss may predict two different flow patterns in the downward pipe for the same production conditions:

- In green, a dispersed flow pattern (the liquid transports gas bubbles) is predicted. The pressure at the bottom of the riser in this case is equal to 340 bar. This is enough to ensure production over the entire remaining line (estimated pressure loss is 10 bar on the horizontal pipe and 320 bar on the upward riser).
- In red, an annular flow is simulated (the liquid wets the entire wall). In this case the pressure obtained at the bottom of the riser is insufficient to ensure production up to the process unit.

This example clearly shows that incorrect prediction of the flow patterns and then of the pressure loss will have dramatic effects on the system's production rate.

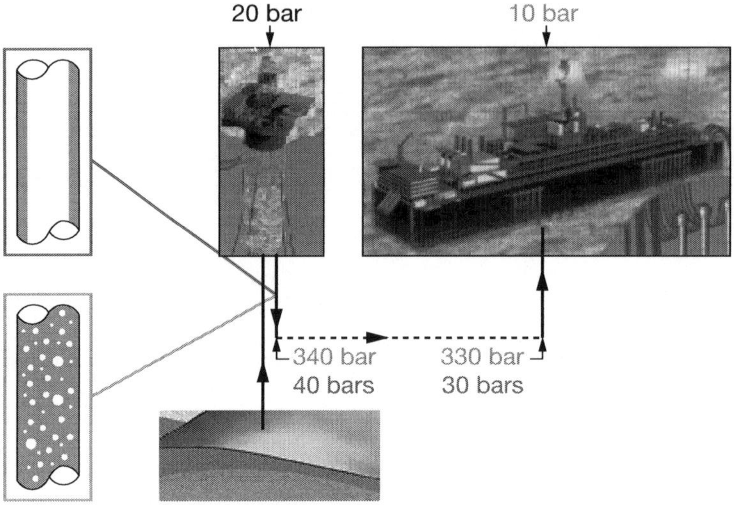

Figure 1.1

Example of export line

2) To illustrate the need for production activation, such as pumping, when natural flow is impossible, we have simulated a 20" multiphase pipeline connecting a well head platform to a production platform. [de Salis *et al.*, 1999].

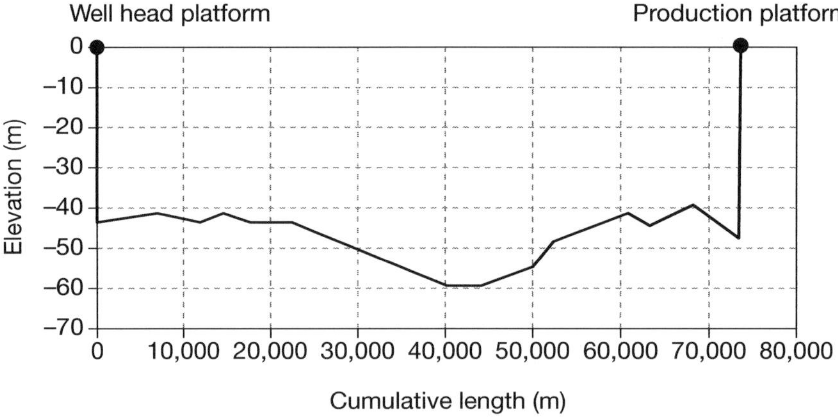

Two problems arose:

- Depending on the liquid and gas flow-rates, the multiphase flow regime in the pipeline can be unstable, owing to the transient severe slugging phenomenon.
- Due to the increase in pressure drop when transporting multiphase flow with a larger liquid stream, the outlet pressure must not be too low to feed the export compressors.

These problems can be solved by installing a multiphase pump to boost the required flow production while maintaining satisfactory pressure levels at pipeline inlet and outlet. Independently of its location, the multiphase pump can be used either to reduce the pipeline inlet pressure or to increase the pipeline outlet pressure, as required by the operator. To study the interaction of the multiphase pump with its downstream or upstream multiphase pipe, numerical simulations of pump locations and production scenarii are run. The conclusion of the study is that in natural production conditions, when severe slugging occurs, the multiphase pump suppresses it and that the multiphase production system can be optimised by adjusting the various parameters (flow production, pump speed, pipeline equilibrium pressure).

This example illustrating some of the problems that the oil and gas industry may encounter, shows the need for multiphase flow prediction in order to detect and solve a production problem or size an item of equipment. To predict the evolution of this kind of flow, thermodynamic and hydrodynamic modelling is necessary.

This chapter presents the different kinds of flow patterns that exist in multiphase flow and the corresponding hydrodynamic, thermodynamic and thermal modelling.

1.2 FLOW PATTERN DESCRIPTION

The mixture transported in the production pipe may be composed of different fluids such as gases, liquids and solids. If we consider a purely hydrocarbon composition for the reservoir fluid, the existing phases are gas (vapour) and liquid. Their compositions change with regard to the pressure and temperature values and are calculated by thermodynamic correlation or flash. However, the fluid reservoir is actually composed of hydrocarbons and other components such as H_2S, CO_2 and water. The presence of water is now unavoidable, as fields are mature or water injection is necessary to help with production. This leads to new problems in hydrodynamic prediction and flow assurance: being able to predict the water evolution in a multiphase mixture is important for corrosion prediction and for the hydrate risk study.

Hydrates are clathrate type crystals in which cages of water molecules are stabilised by host molecules [Sinquin *et al.*, 2004].

Unlike ice crystals, gas hydrate crystals are able to form at temperatures higher than 0°C as soon as the pressure is higher than a few tens of bars. Conditions promoting hydrate formation are high pressure (typically > 30 bar) and low temperature (typically < 20°C). Precise conditions in terms of pressure and temperature depend on the composition of the fluids. Hydrate formation can occur for all the fluids produced if required P-T conditions are reached: natural gas, gas condensate and crude with associated gas, with condensed or formation water.

Formation of hydrate particles generally leads to the formation of solid plugs, blocking pipelines and leading to shutdown of production facilities.

To anticipate and solve potential production problems related to hydrate blockages, operators have at their disposal of tools and resources in the following areas:

- Prediction methods: essentially involve performing thermodynamic calculations that enable dissociation curve to be determined for hydrates.

- Prevention: The most common methods presently used or foreseen by operators to maintain pressure and temperature conditions outside the hydrate formation region are insulation and the injection of thermodynamic inhibitors. A new option would be the injection of the so-called "Low Dosage Hydrate Inhibitors". Some of them act by delaying hydrate nucleation and by slowing down crystal growth, while others their transport in suspension feasible by preventing hydrate deposition and the formation of large aggregates.
- Remediation: the method so far successfully implemented by operators is two-sided depressurization

Finally, during the production of gas and liquid, the flow variation may create instabilities leading to the break-up of the perforation wall. This phenomenon leads to the presence of sand in the production fluid. To optimise hydrocarbon production by controlling or preventing sand production, the oil company carried out "Sand Management" studies. In this case the well is not prepared with particular conditions such as a gravel pack and the index of production is better, but good knowledge of the quantity of sand produced is required in order to to size the surface separators and prevent pipe erosion. The models developed are flow pattern dependent.

As we will see in this chapter the flow patterns differ according to the nature of the existing phases. To determine them, visual methods based on conventional photography [Hewitt *et al.*, 1969], high-speed photography [Trallero, 1995] and X-Ray photography [Derbyshire *et al.*, 1969] have been used. Other methods, based on pressure measurement [Lin *et al.*, 1987; Tutu, 1982], photon attenuation [Jones *et al.*, 1975], or electrical conductivity [Barnea *et al.*, 1980; Flores, 1997] have also been used.

1.2.1 Gas-Liquid Flow

Using a sufficient number of adjectives it is possible to describe a large series of flow states [Duns *et al.*, 1963; Hubbard *et al.*, 1966; Wallis, 1969; Govier *et al.*, 1972; Bergles *et al.*, 1981]. However, this would not be much use for modelling purposes. This explains why a large number of configurations are generally grouped into three families:

- **Separated flow**: because of gravity, the gas and the liquid flow separately in the pipe. Depending on the interface structure we can define stratified smooth (**SM**) (linear interface) flow observed for low flow-rates (Fig. 1.2). When the rate is increased, the flow becomes stratified wavy (**SW**): some waves appear on the interface. When the gas velocity is very high, or in the case of an inclined pipe, the liquid forms a continuous film on the wall and the gas flows at the centre of the pipe. This is referred to as annular flow (**AN**). Some drops of liquid can be dragged out of the film and transported by the gas. This is called annular dispersed or Mist Flow (**AD**) or wispy annular when there is a swarm of drops.
- **Bubbly flow (B)**: For high liquid flow rate and low gas flow rate the gas phase is transported by the liquid in the form of bubbles. Below a diameter of 1 mm the bubbles are generally considered to be spherical (Fig. 1.3).

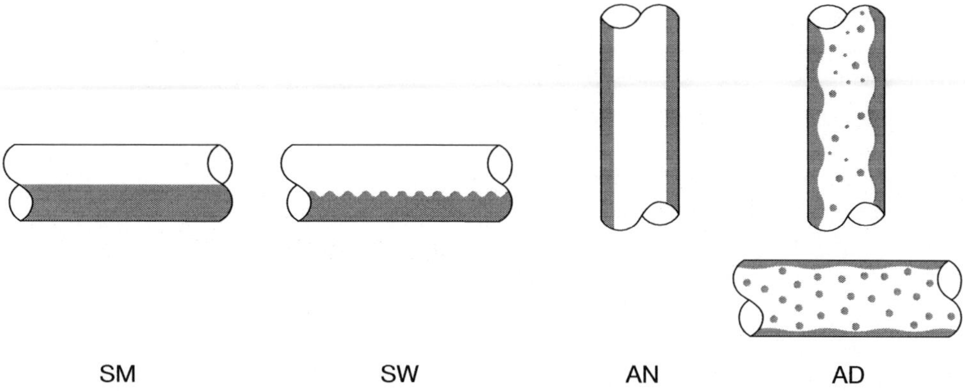

Figure 1.2

Stratified flow patterns

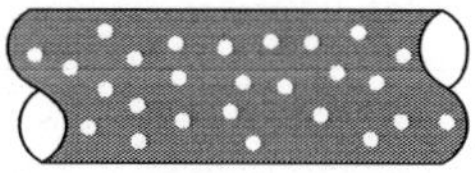

Figure 1.3

Bubbly flow pattern (B)

- **Slug flow (SL)**: Between the above two configurations, there is a transitional flow pattern (Fig. 1.4). It consists of Taylor bubbles (long bubbles) separated by liquid slugs. These slugs can contain small bubbles dragged from the tail of the Taylor bubble.

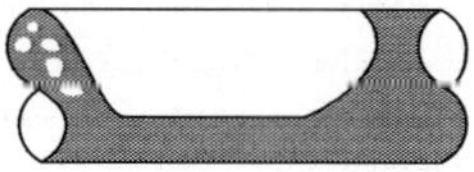

Figure 1.4

Slug flow pattern with aerated slug body (SL)

1.2.2 Liquid-Liquid Flow

The two-phase liquid-liquid flow is far more complex than the previous case because the densities of both phases are of the same order. Trallero [1995] proposed a new classification based on published and acquired data for horizontal flow. For the dispersed case each flow pattern can exist either in oil dominated (**W/O**, water in oil) or in water dominated (**O/W**, oil in water) depending on the status of the continuous phase. The process during which an oil-water mixture changes from O/W to W/O is called phase inversion. Flores [1997] completed Trallero's study for the vertical and deviated pipes. The major difference in the flow patterns is the presence of counter-current in the pipe.

- **Separated flow**: Because of gravity, the two phases tend to separate in the pipe. Depending on the interface configuration three cases can be detected:
  - Stratified smooth **(SM)**: the interface is linear.
  - Stratified wavy **(SW)**: some waves deform the interface structure.
  - Mixing zone (**SMix**): at increased flow-rate, the deformation of the interface becomes more pronounced and a dispersion zone appears between the two layers as shown on Figure 1.5. This flow pattern may also be obtained by increasing the fraction of the dispersed phase. In this case the droplets coalesce, to create the second "pure" phase, a layer of dispersion remaining at the interface between the oil and water phase.

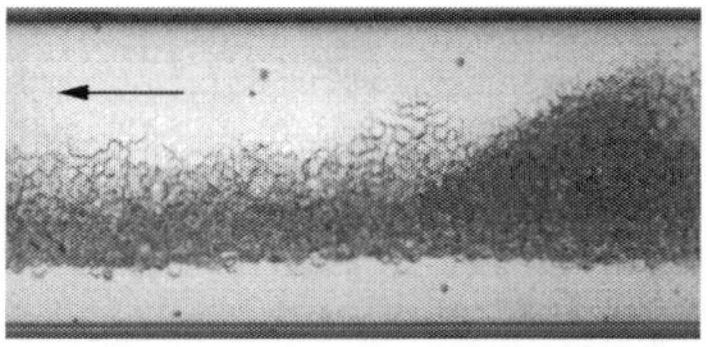

Figure 1.5

Stratified flow with a mixing zone at the interface [Conan, 2007]

- **Dispersed flow**: Depending on the fluid characteristics and the input flow rates, different kinds of dispersed flow can be detected depending on the size of the droplets (Fig. 1.7). To simplify the presentation of the flow patterns, we here only present the water dominated flow patterns, with an equivalent oil dominated flow existing for each of them:
  - Oil in water very fine dispersion (**O/W–VFDis**): Regular distribution of very small oil droplets over the entire pipe cross-section. No slip between the two phases.
  - Oil in water **(O/W)**: The shape and the distribution of the oil droplets are irregular and depend on the fluid characteristics and the inlet flow rates.
  - Oil in water Churn flow **(O/W–CF)**: Presence of very large oil droplets.
  - Oil in water and water **(O/W & W)**: Dispersion of oil in water flowing over a water layer as shown in Figure 1.6.

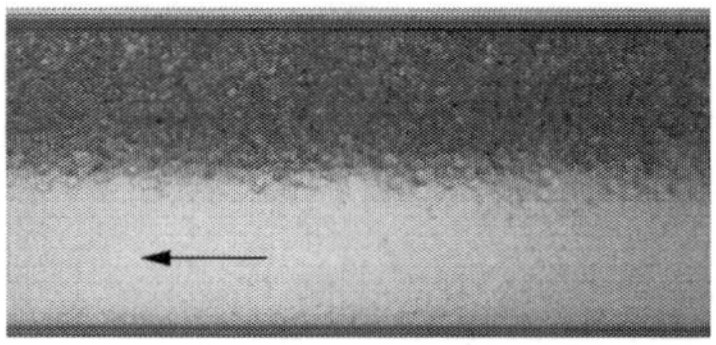

Figure 1.6

Oil in water dispersion flowing over a pure water layer [Conan, 2007]

- Oil in water and water in oil (**O/W & W/O**): The two types of dispersions coexist.

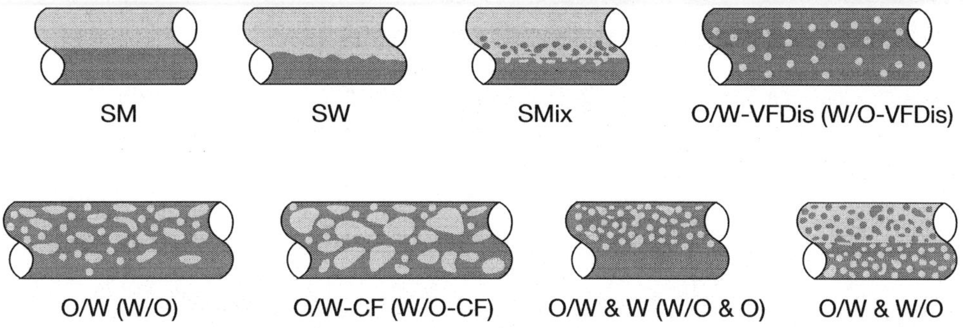

Figure 1.7

Liquid-liquid flow patterns

1.2.3 Gas-Liquid-Liquid Flow

Depending on the flow rate of the gas and the liquids, mixing of the gas-liquid and liquid-liquid flow patterns appears (Fig. 1.8). The three phase flow configurations can be classified according to either the liquid-liquid state or the gas-liquid state. A more detailed description of these flows is given by [Açikgöz *et al.*, 1992; Pan, 1996; Odozi, 2000]. For the gas/liquid-liquid system, the description is the same as for the two-phase gas-liquid flow. For the liquid-liquid part, the flow is simply described as dispersed or stratified. The innovation lies in the mixing of the gas-liquid and liquid-liquid flow configurations.

- **Separated liquid-liquid flow**: gravitational separation appears between the two liquids. Depending on the fluid characteristics and the flow history, the interface between the two liquids can be smooth, wavy (S–StWa) or composed of oil or water droplets. For relatively low gas flow rates a gas-liquid slug flow (S–Sl) is detected, with the Taylor bubble flowing in the lighter liquid. If we increase the gas flow rate, the long bubbles disappear and complete stratification of the three phases takes place. The gas-liquid interface can be smooth or wavy.
- **Dispersed liquid-liquid flow**: If we increase the mixing of the two liquids, the liquid-liquid interface breaks and dispersion is detected. Depending on the oil/water cut, the dispersion can be oil or water dominated. The gas-liquid configuration is the same as previously described (OinW–Sl, OinW–StWa).
- **Liquid-liquid mixed flow**: Odozi observed a more complex liquid-liquid inter-action in the gas-liquid slug flow. In the slug body the two liquids are well mixed whereas in the film region, between the slugs, there is some stratification in the liquid layer. This stratification is possible because of the low slug frequency. Once again the liquid mixture in the slug body can be continuous oil or water.

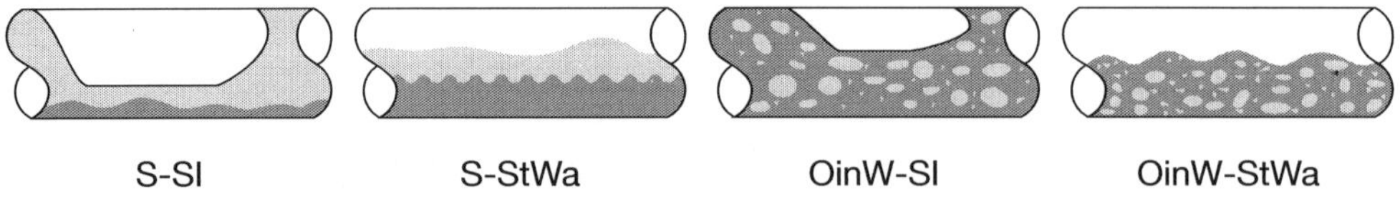

Figure 1.8

Examples of Gas-liquid-liquid flow pattern

1.2.4 Solid Suspensions

Depending on the nature of the transporting phase (gas or liquid), the flow patterns will be different:

- **Gas-particles**: In a horizontal flow with a high gas velocity, turbulent mixing maintains a homogeneous state (Fig. 1.9). With a reduction of the gas velocity, the particles begin to settle out and collect at the bottom of the pipe. This is called the dune flow. Further reducing the velocity leads to a slug flow, until at a certain gas velocity, the particles completely fill the pipe and lead to a packed bed. When this last flow pattern appears, pneumatic transport ceases.

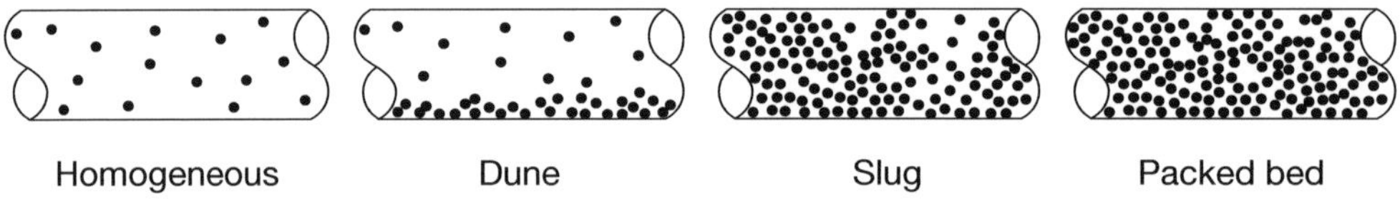

Figure 1.9

Horizontal pneumatic (gas/solid) transport

- **Liquid-particles**: The transport of particles (mud, sand, hydrates) by a liquid is identified as slurry flow. Once again, depending on the characteristics of the particles and the fluid, and on the velocity of the liquid, three flow patterns are identified (Fig. 1.10). At high liquid velocity, turbulence maintains a homogeneous flow. As the velocity decreases, the particles settle out leading to a moving bed regime, before ending up with a stationary bed.

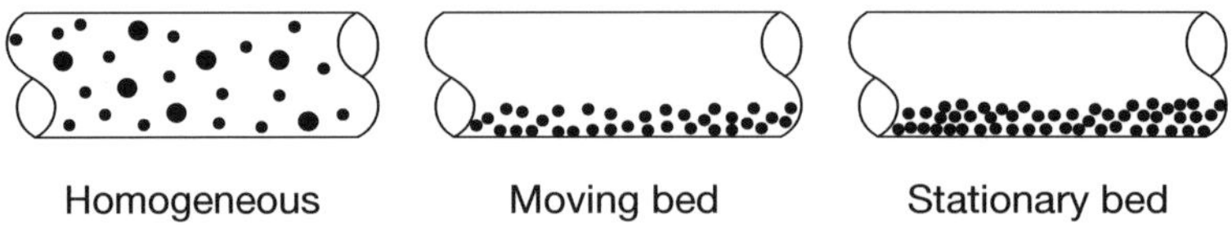

Figure 1.10

Liquid/Solid transport

1.2.5 Flow Pattern Transitions

By varying the inlet flow rates, the fluid properties and the geometrical configurations, initial maps classifying the flow patterns were experimentally drawn up over a wide range of coordinates. Generally the maps are presented in terms of the flow rates. It is clear that the transition lines are highly dependent on all these parameters. To illustrate this sensitivity we have represented on the flow map in Fig. 1.11 a case study of a condensate/natural gas at 10 bar in a 3" vertical pipe. Fig. 1.12 represents the flow map for the same system in a 6" pipe; the slug region is smaller and mist flow appears at a gas velocity of 10 m/s. On Fig. 1.13 the angle of inclination has been changed and a degree of stratified wavy flow appears, while on Fig. 1.14 the pressure has been increased, changing the physical properties of the fluids and the position of the transitions. On these four maps the sensitivity of the transition lines regarding the pipe diameter, pipe angle of inclination and pressure is important.

Secondly, the need for predictive tools drove the researchers to write models for the transition laws dependent on physical or geometrical parameters

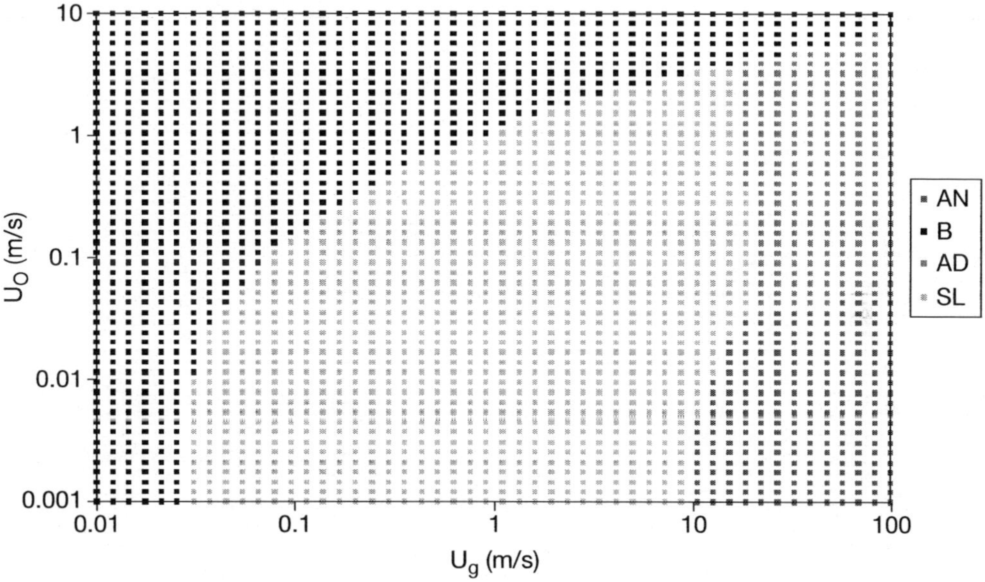

Figure 1.11

Condensate, natural gas flow regime map. 10 bar, θ = 90°, D = 3"

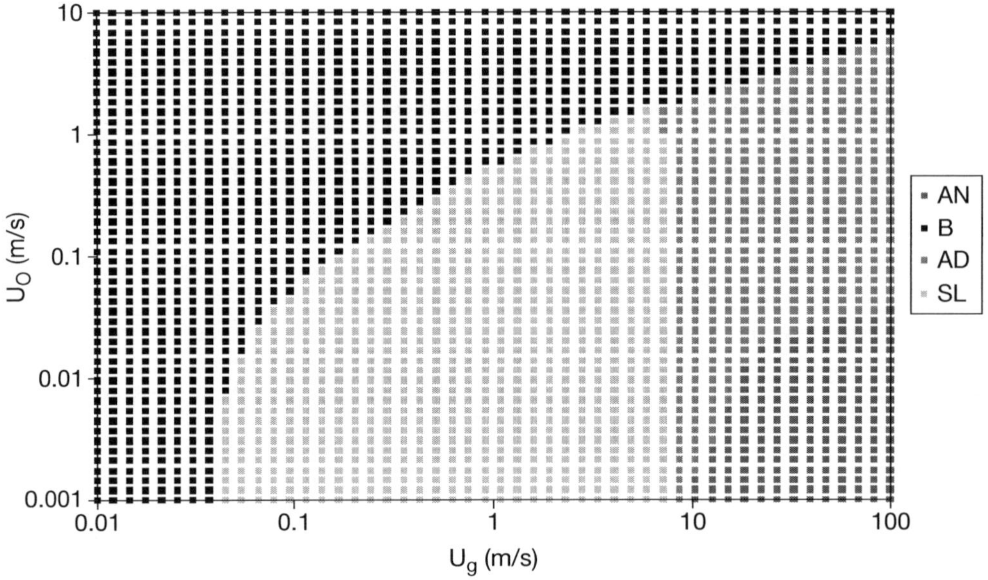

Figure 1.12

Condensate, natural gas flow regime map. 10 bar, θ = 90°, D = 6"

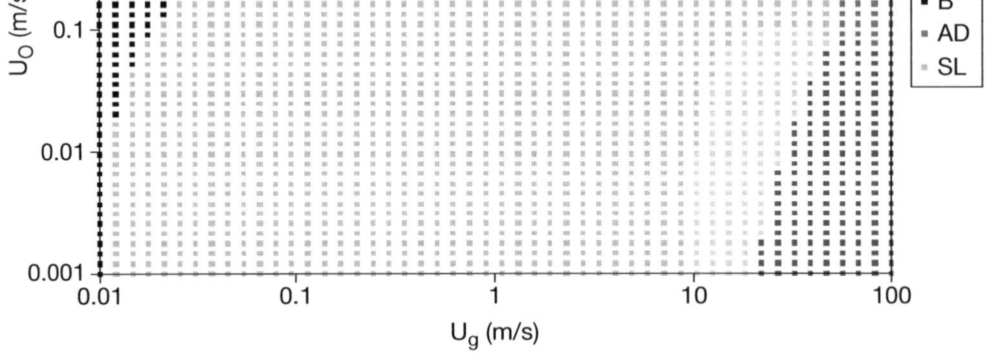

Figure 1.13

Condensate, natural gas flow regime map. 10 bar, θ = 15°, D = 3"

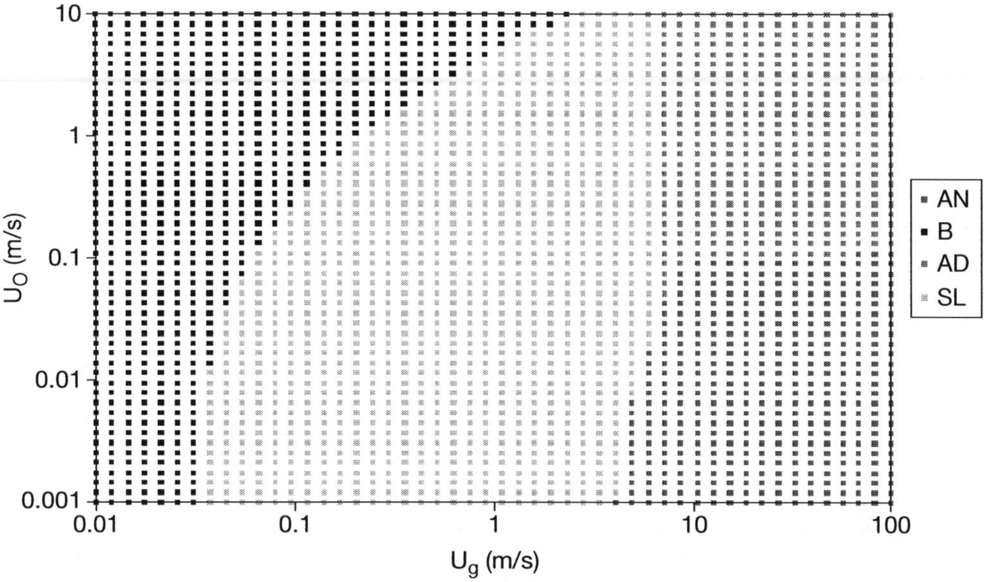

Figure 1.14

Condensate, natural gas flow regime map. 40 bar, θ = 90°, D = 3"

- **Gas-Liquid Flow**

Baker [1954] proposes the earliest complete study (stratified smooth and wavy, annular, dispersed, bubble and slug flow) for horizontal pipes. More recently, Taitel *et al.* [1976] presented a theoretical model to predict the transition for horizontal and slightly inclined pipes. This model depends on gas and liquid mass flow rates, fluid properties, pipe diameter and the angle of inclination from the horizontal. It is fully predictive as no flow regime data are used in its development. Five flow patterns (stratified smooth, stratified wavy, annular, intermittent, dispersed bubble) are considered. Shoham [1983] conducted an experimental study for horizontal, upward and downward inclined and vertical flow for a 2.54 cm and 5.1 cm pipe. His study is probably the most complete as he uses a complete 360° range for the angle of inclination and uses two different diameters. He compares his experimental results with the Taitel and Duckler model and proposes an extension of this model.

For concurrent upflow, maps based largely on experimental work are suggested by Duns *et al.* [1963], Wallis [1969], Hewitt *et al.* [1969] and Govier *et al.* [1972]. Griffits *et al.* [1961], Oshinowo *et al.* [1974], Spedding *et al.* [1980] and Taitel *et al.* [1980] present maps based on physical modelling. For the cocurrent downflow very few studies exist: Golan *et al.* [1969] who made a study of an inverted U was the first to produce a map for downflow. This was followed by Martin [1973], Yamazaki *et al.* [1979], Spedding *et al.* [1980] who focus on the transition from bubble to slug flow and Crawford *et al.* [1985] who reports data collected with 2.5 cm and 3.8 cm tubes. More recently Sekogushi *et al.* [1996] and Hajiloo *et al.* [2001] have published some experimental data.

- **Liquid-Liquid Flow**

Brauner *et al.* [1992] proposes a horizontal flow model of transitional boundaries based on mechanistic models. Unfortunately, a limited amount of experimental data was found to validate their model. Brauner [2001] presents a more complete model to predict the boundaries of the dispersed flow pattern that takes account of fluid properties, tube diameter and angle of inclination. Trallero [1995] using his own experimental databank, bases his study of transition on two main flow patterns: separated phases and dispersed flow. He shows that this transition must be addressed with the complete two-fluid model. Comparisons with published data also correspond closely. His study was completed by Flores [1997] for vertical and diverted pipes. He proposes a mechanistic model to predict the flow pattern transition governing the transitions between the dispersed flow and the very fine dispersed and Churn flow respectively. The prediction of phase inversion has been covered mostly in "static" condition (mixers). Only a few studies have been carried out using pipes. Decarre *et al.* [1997] presents a review of the existing literature and a model to predict the critical fraction of dispersion at which the nature of the dispersed phase changes. Relatively good correspondence is obtained regarding data found in the literature. More recently, Brauner *et al.* [2002] developed a model applicable to very dense dispersions.

- **Gas-Liquid-Liquid Flow**

Most of the studies dealing with three-phase flow pattern maps are experimental. Malinowski [1975] for horizontal flow and Shean [1976] for vertical flow are among the earliest researchers to have carried out work on three-phase flow maps. Both use a two-phase gas-liquid map to present their results. A much more detailed description of the flow patterns is introduced by Acikgoz *et al.* [1992]: the classification of the flow is given regarding the gas-liquid state and the liquid-liquid state. Ten flow patterns are identified and the map is given in terms of water and gas superficial velocities for a fixed oil superficial velocity. In fact the difficulty of these maps is to represent the three-dimensional entry on a two-dimensional map. Using the same classification, Pan [1996] detects two other flow patterns. By comparison with existing two and three-phase maps he concludes that none of the existing maps satisfactorily predicts his data acquired at atmospheric and high pressure. Odozi [2000] presents a study showing the effect of the pressure and the pipe inclination angle on the position of the transitions. His conclusion regarding the use of other three-phase maps to classify his data is the same as Pan's. This confirms that the transitions depend on both the geometry and the fluid characteristics.

1.3 PHYSICAL MODELLING

1.3.1 Basic State Equations

For a pipe of constant section A and axis (oz) we note $\mathbf{n_z}$ the unit vector on the axis. Phase k, materialised by the volume V_k can be in contact with others phases. Looking at a section of pipe containing this phase, we note $\mathbf{n}_{kj}$ the normal unit vector, directed outwards, on the interface between phases k and j as presented in Figure 1.15:

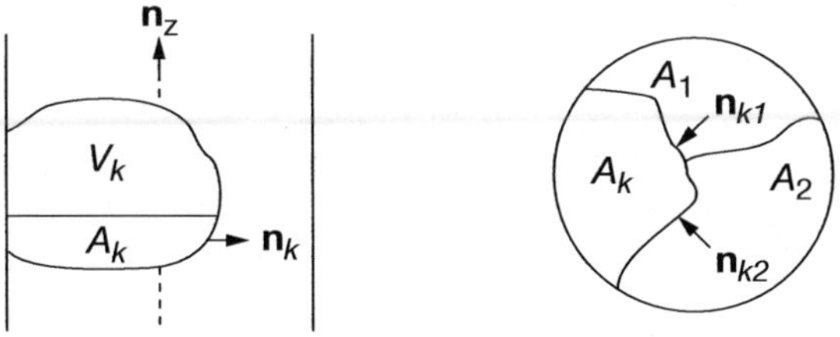

Figure 1.15

Pipe section

Integral balance laws, written for a fixed volume containing all the phases, leads to the local instantaneous phase equations. They can be defined by a general form of these equations:

$$\frac{\partial}{\partial t}\left(\rho_k f_k\right) + \nabla\left(\rho_k \mathbf{u}_k f_k\right) = \nabla.\mathbf{J}_k + b_k \tag{1.1}$$

With ρ_k and $\mathbf{u_k}$ the density and velocity of phase k respectively. The density will be assumed to be constant in our problem. f_k, $\mathbf{J}_k$ and b_k are defined by Table 1.1. The local instantaneous jump equation is:

$$\sum_k \sum_{j\neq k}\left(\rho_k f_k\left(\mathbf{u}_{kj} - \mathbf{u}_k\right).\mathbf{n}_{kj} + \mathbf{n}_{kj}.\mathbf{J}_k\right) = 0 \tag{1.2}$$

Table 1.1 Equivalence table for the mass and momentum balance and the energy equations

	f_k	$\mathbf{J}_k$	b_k
Mass balance	1	0	0
Momentum balance	$\mathbf{u}_k$	$\boldsymbol{\tau}_k \quad p_k\mathbf{I}$	$\rho_k\mathbf{g}$
Total energy	$e_k + \mathbf{u}_k^2/2$	$\left(\boldsymbol{\tau}_k - p_k\mathbf{I}\right)\mathbf{u}_k - \mathbf{q}_k$	$\rho_k\mathbf{g}\,\mathbf{u}_k$

Then, using space/time or time/space average methods, the equations commonly used to solve the practical problems of multiphase flow in pipelines are obtained. These methods are well described by Ishii [1975] or Bergles *et al.* [1981]. The conservation equations averaged over the section can be summarised by the following equation:

$$\begin{aligned}\frac{\partial}{\partial t}\left(R_k\rho_k F_k\right) + \frac{\partial}{\partial z}\left(R_k\rho_k U_k F_k\right) &= \frac{\partial}{\partial z}\left(R_k J_{kz}\right) + R_k B_k + J_k^w \\ &\quad - \frac{\partial}{\partial z}\left(R_k\rho_k\left\langle \tilde{f}_k \tilde{\mathbf{u}}_k \mathbf{n}_z\right\rangle\right) + \sum_{j\neq k}\left(G_{kj}^{\mathrm{i}} + J_{kj}^{\mathrm{i}}\right)\end{aligned} \tag{1.3}$$

With $\mathbf{u}_k = \mathbf{U}_k + \tilde{\mathbf{u}}_k$ and $U_k = \mathbf{U}_k\ \mathbf{n}_z$, the average value $B_k = <b_k>$ defined as for $F_k = \langle f_k \rangle = \frac{1}{A_k} \iint_{A_k(z,t)} f_k(z,t)\, dS$, $J_{kz} = \langle \mathbf{J}_k \mathbf{n}_z \rangle$ being the axial component of the quantity $\mathbf{J}_k$, $J_k^W = \frac{1}{A} \int_{C_w(z,t)} \mathbf{J}_k\, \mathbf{n}_\mathbf{w}\, dC$ its wall contribution and $\sum_{j \neq k} J_{kj}^{\mathrm{i}} = \sum_{j \neq k} \frac{1}{A} \int_{C_{kj}(z,t)} \mathbf{J}_k\, \mathbf{n}_{kj}\, dC$ its interfacial contribution. The interfacial parameters associated with the mass transfer are defined as $\sum_{j \neq k} G_{kj}^{\mathrm{i}} = \sum_{j \neq k} \frac{1}{A} \int_{C_{kj}(z,t)} \rho_k f_k \left(\mathbf{u}_{kj} - \mathbf{u}_k \right) \mathbf{n}_{kj}\, dC$. Once the general form of the jump equation is:

$$\sum_k \sum_{j \neq k} \left(G_{kj}^{\mathrm{i}} + J_{kj}^{\mathrm{i}} \right) = 0 \tag{1.4}$$

Generally, when dealing with the problem of transport in pipelines, the above equations can be simplified assuming an incompressible, fully developed flow. Assuming no mass transfer, the basic equations for the one-dimensional case are:

$$\frac{\partial \varepsilon_k \rho_k}{\partial t} + \frac{\partial \varepsilon_k \rho_k U_k}{\partial z} = 0 \tag{1.5}$$

$$\frac{\partial \varepsilon_k \rho_k U_k}{\partial t} + \varepsilon_k \frac{\partial P}{\partial z} = \frac{P_{\mathrm{w}k}}{A} \tau_{\mathrm{w}k} + \frac{P_{\mathrm{i}k}}{A} \tau_{\mathrm{i}k} - \varepsilon_k \rho_k g \sin\theta \tag{1.6}$$

$$\frac{\partial}{\partial t}\left(\sum_k \rho_k \varepsilon_k \left(E_k + \frac{U_k^2}{2} \right) \right) + \frac{\partial}{\partial z}\left(\sum_k \rho_k \varepsilon_k U_k \left(E_k + \frac{U_k^2}{2} \right) \right) = -\frac{\partial}{\partial z}\left(P \sum_k \varepsilon_k U_k \right) + \frac{Q_{in}^W}{A} - g \sin\theta \sum_k \varepsilon_k \rho_k \tag{1.7}$$

For two-phase flow problems this model is usually called the two-fluid model. Equations 1.5 to 1.7 are used to solve unsteady problems. Dealing with one-dimensional steady flow, the time-dependent terms are removed from these equations.

1.3.2 Drift Flux Model

The drift flux model is a model based on the volumetric flux of a component relative to a surface moving at the average velocity, commonly named drift flux, which is proportional to the relative velocity (difference between the mean velocities of two phases). This model, which is well described by Wallis [1969] rapidly offers a solution for multiphase flow problems.

When dealing with a two-phase flow system and given that $j_k = \varepsilon_k U_k$, $\sum_{k=1,2} \varepsilon_k = 1$ and $V_{rel} = U_G - U_L$, a second-order equation depending on R_G can be written:

$V_{rel}\varepsilon_G^2 - \left(j_G + j_L + V_{rel}\right)\varepsilon_G + j_G = 0$. If this equation can accept solutions, depending on the nature of the flow pattern, the slip velocity between the two phases may be estimated by different models and the void fraction is then calculated using:

$$\varepsilon_G = \frac{j_G + j_L + V_{rel} \pm \sqrt{\left(j_G + j_L + V_{rel}\right)^2 - 4V_{rel}j_G}}{2V_{rel}}$$

1.3.3 Hydrodynamic Closure Laws for the Transportation Equations

To solve the problem using simplified state equations 1.5 to 1.7, some closure laws are needed. These equations will depend on the flow pattern.

1.3.3.1 Stratified Flow

The parameters to be defined are the shear stresses at the wall and at the interfaces. Several studies have been carried out for the two-phase gas-liquid flow while only a few articles present original models for liquid-liquid and gas-liquid-liquid flows.

- **Gas-Liquid Flow**

To obtain the pressure drop and the phase fractions, the momentum equation is generally written for the mixture:

$$\frac{\partial}{\partial t}\left(\rho_G U_G - \rho_L U_L\right) = \frac{P_{wG}}{A\varepsilon_G}\tau_{wG} - \frac{P_{wL}}{A\varepsilon_L}\tau_{wL} + \frac{P_{GL}}{A\varepsilon_L\varepsilon_G}\tau_{GL} - \left(\rho_L - \rho_G\right)g\sin\theta \quad (1.8)$$

In order to solve this equation, the wall and interfacial shear stresses are expressed as a function of the other parameters. To reduce the number of parameters, friction factors are introduced:

$$f_{wk} = \frac{-\tau_{wk}}{\frac{1}{2}\rho_k\left|U_k\right|U_k} \quad \text{and} \quad f_{GL} = \frac{-\tau_{GL}}{\frac{1}{2}\rho_G\left|U_G - U_L\right|\left(U_G - U_L\right)} \quad (1.9)$$

The definition of the wall friction factors is the same as that used for single-phase flow. Bird *et al.* [1960] explain why it depends on the Reynolds number and the roughness of the pipe. For laminar flow, power laws are usually applied. For example, the Hagen-Poiseuille formula applied to the two-phase flow gives: $f_{wk} = 16/\mathrm{Re}_k$ with $\mathrm{Re}_k = Dh_k\left|U_k\right|/\nu_k$

and $Dh_{\mathrm{G}} = \dfrac{4A\varepsilon_{\mathrm{G}}}{P_{\mathrm{wG}} + P_{\mathrm{GL}}}$, $Dh_{\mathrm{L}} = \dfrac{4A\varepsilon_{\mathrm{L}}}{P_{\mathrm{wL}}}$, the interface being considered as a wall by the gas and a free surface by the liquid. For turbulent flow, the Blasius power law $f_{wk} = 0.079/\mathrm{Re}_k^{0.25}$ can be used for smooth pipes, while for rough pipes, Colebrook's formula can be extended to two-phase flow: $\dfrac{1}{\sqrt{f_{wk}}} = 3.48 - 4\log\left(\dfrac{2k_{\mathrm{w}}}{Dh_k} + \dfrac{9.35}{\mathrm{Re}_k\sqrt{f_{wk}}}\right)$.

For a smooth interface, expression of most of the friction factors depends on the wall friction factor: Crouzier [1978], Mouly [1979], Andritsos *et al.* [1987]. Kowalski [1987] (Eq. 1.10) introduces a relation depending only on the Reynolds numbers and the liquid holdup.

$$f_{\mathrm{GL}} = 0.96\left(\frac{j_{\mathrm{G}}D}{\nu_{\mathrm{G}}}\right)^{-0.52} \tag{1.10}$$

Kowalski [1987] also proposed a relation for a wavy interface (Eq. 1.11):

$$f_{\mathrm{GL}} = 7.5\times10^{-5}\varepsilon_{\mathrm{L}}^{-0.25}\mathrm{Re}_{\mathrm{G}}^{-0.3}\mathrm{Re}_{\mathrm{L}}^{0.83} \tag{1.11}$$

This last equation is valid for: $22600 \le \mathrm{Re}_{\mathrm{G}} \le 430600$ and $8800 \le \mathrm{Re}_{\mathrm{L}} \le 47800$.

Charnock [1955] shows that the surface roughness increases with the friction velocity. Cohen *et al.* [1986] demonstrate that wave amplitude also has an influence on roughness. In fact, none of the various areas of research has shown itself to be better than the others and the wave mechanism in shallow water is not yet completely resolved.

- **Liquid-Liquid Flow**

The problem is more complex than for gas-liquid because of the nature of the interface. Most of the time, an oil-water mixture separates the oil and the water layers. The two parameters to be modelled are the wall and interfacial shear stresses. Most of the time the authors use gas-liquid correlation. Lovick *et al.* [2004] presented comparisons of their experimental data with such a model. They show that models of this kind need to be corrected to include the effect of the interface shape and the entrainement of one phase into the other. Conan [2006] has conducted a complete study on stratified mix flow using a PIV system and a high-speed camera. The velocity and turbulence profile obtained over the section have been used to determine the wall shear stress in the pure water phase and the dispersion layer.

- **Gas-Liquid-Liquid Flow**

Taitel *et al.* [1995] develops a three-fluid model using very simple friction laws for the interfaces:

$$f_{\mathrm{GL}} = \max\left(0.014, f_G\right) \tag{1.12}$$

$$f_{\mathrm{OW}} = \max\left(0.014, f_O\right) \tag{1.13}$$

No original work has been done regarding the closure laws for the gas-liquid and the liquid-liquid interface. This is mostly due to the lack of precise experimental data and the fact that liquid-liquid flow is poorly understood.

1.3.3.2 Dispersed Flow

- **Gas-Liquid Flow**

The closure law used for this configuration is generally the gas velocity obtained by the Zuber & Findlay [1965] method:

$$U_G = C_0 U_M + f(R_G) U_\infty . \tag{1.14}$$

This law takes account of the variation in bubble concentration through the flow distribution parameter C_0 (its value lies between 1.0 and 1.5 with a usual value of 1.2). U_∞ is the terminal velocity of a bubble and $f(R_G)$ is a function integrating the influence of the void fraction. Wallis [1969] gives a review of terminal velocities depending on the flow conditions. For example, for the smallest bubbles, considered to be spherical, Stokes' solution can be used: $U_\infty = \dfrac{d^2 g(\rho_L - \rho_G)}{18\mu_L}$ while for large bubbles, the effects of surface tension and viscosity are negligible and the equation of Davies *et al.* [1950]: $U_\infty = \sqrt{g\dfrac{d}{2}}$ will be used. For intermediate bubble diameters, a large number of correlations enables Peebles *et al.* [1953] to present velocities applicable for a wide range of Re_G values.

Most terminal velocity definitions include the bubble diameter. This diameter depends on the bubble formation mechanism and the flow conditions. In the case of formation by Taylor instability the diameter will be defined by: $d = 4\sqrt{\dfrac{\sigma_{GL}}{g(\rho_L - \rho_G)}}$ while to take account of shear stresses, based on the study by Hinze [1955], Sevik *et al.* [1973] proposes $d = 1.15\left(\dfrac{\sigma_{GL}}{\rho_L}\right)^{3/5} \varepsilon_{dis}^{-2/5}$.

- **Liquid-Liquid Flow**

Depending on droplet size, shape and distribution, different cases are defined. If dispersion is "homogeneous", in other words comprising uniform spherical droplets homogeneously distributed over the pipe section, the drift flux between the continuous phase and the droplets may be considered null. In this case the homogeneous model can be applied with good results. If the droplets are bigger and have a less spherical shape, a slip velocity appears. In this case a two-fluid model of the slip velocity can be used.

- The homogeneous model: this consists in considering the mixture as a single phase. By correctly defining the physical properties of this phase it is possible to apply single-phase correlations to determine the pressure drop. The density of the mixture is proportional to the quantity of each phase present:

$$\rho_M = \sum_k \rho_k \alpha_k \tag{1.15}$$

The problem with this method is to correctly estimate the apparent (or relative) viscosity μ_M. In the literature, there is a wide range of correlation, directly dependant on the fluid system in question. Pal [2001] gives a good review of relative viscosity expressions and proposes a new equation integrating the viscosity ratio:

$$\eta_r \left(\frac{2\eta_r + 5\eta_d / \eta_c}{2 + 5\eta_d / \eta_c} \right)^{3/2} = \left(1 - \frac{\varepsilon_d}{0.637} \right)^{-1.59} \tag{1.16}$$

- Slip velocity definition:
  Flores [1997] extended the drift flux model developed by Zuber *et al.* for gas-liquid (Eq. 1.12). The distribution parameter is equal to 1 if the droplet concentration is uniform. If the droplets are concentrated at the centre of the pipe it is greater than 1. If they are concentrated at the wall it is less than 1. The terminal velocity depends on the nature of the continuous phase.

For the case of a more complex flow pattern such as the O/W & W presented in Fig. 1.6, Conan [2007] shows that the pressure drop is mainly imposed by the dispersion layer. A PIV technique was used to determine the wall shear stress.

- **Gas-Liquid-Liquid Flow**

The literature in fact reports no flow pattern for which the gas is a dispersed phase, while dispersions of oil in water or water in oil have been reported. The laws used to deal with these dispersions are the same as that used for the two-phase gas-liquid or liquid-liquid flow.

1.3.3.3 Slug Flow

- **Gas-Liquid Flow**

Hale [2000] presents a detailed review of the existing literature and gives a complete study (experiments and modelling) of slug formation, growth and decay.

To simplify the problem, the flow pattern can be considered as a cell comprising two parts: the bubble and the slug (Fig. 1.16). Brauner and Ullman [2004] take into account a third part, the Taylor bubble wake.

One way of solving the problem is to treat the bubble part as a stratified flow and the slug part as a bubbly flow.

The bubble velocity is generally defined by a Nicklin *et al.* [1962] type of law. Experiments conducted by Hout *et al.* [2002] show that for small pipe diameters (around 2 cm) this kind of law corresponds well for a wide range of pipe inclination angles, while for larger pipe diameters (around 5 cm), a substantial discrepancy appears between predicted and experimental results.

To predict the gas fraction in the slug body, either a correlation or an equation modelling the bubble entrainment can be applied.

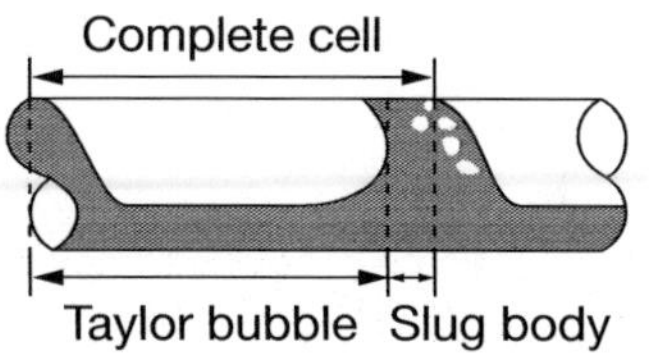

Figure 1.16

Slug flow configuration

Gregory *et al.* [1978] proposed an empirical correlation for horizontal flow: $R_{Lslug} = \left(1 + \left(\frac{U_M}{8.66}\right)^{1.39}\right)^{-1}$. Completing these experimental data with inclined and vertical flows, Malnes [1982] proposes an alternative form taking account of the physical properties of the fluids: $R_{Gslug} = 1 - \frac{U_M}{83\left(g\sigma_{GL}/\rho_L\right)^{0.25} + U_M}$. Ferschneider [1983] later proposes: $R_{Gslug} = \left(1 + \left(\frac{U_M}{\sqrt{gD\left(1-\rho_G/\rho_L\right)}}\frac{Bo^{0.1}}{25}\right)^2\right)^{-2}$ with the Bond number defined by: $Bo = \frac{\left(\rho_L - \rho_G\right)gD^2}{\sigma_{GL}}$. A more mechanistic correlation for horizontal and slightly inclined pipes is proposed by Andreussi *et al.* [1989]:

$$\begin{cases} R_{Gslug} = \dfrac{U_M - U_{Mf}}{\left(U_M - U_{M0}\right)^{\left(1-3\rho_G/\rho_L\right)}} & \text{for } U_M < U_{Mf} \\ R_{Gslug} = 1 & \text{for } U_M \leq U_{Mf} \end{cases}$$

$$\text{with} \begin{cases} U_{Mf} = 2.6\left(1 - 2\left(\dfrac{0.025}{D}\right)^2\right)\sqrt{gD\dfrac{\left(\rho_L - \rho_G\right)}{\rho_L}} \\ U_{M0} = 2400\left(1 - \dfrac{\sin\theta}{3}\right)Bo^{-3/4}\sqrt{gD\dfrac{\left(\rho_L - \rho_G\right)}{\rho_L}} \end{cases}$$

Where U_{Mf} corresponds to the velocity below which no bubbles are present in the slug.

The other way to define the gas fraction in the slug body is to determine the rate of gas entrainment from the rear of a Taylor bubble. Of the few studies done on this subject we should mention Riiser *et al.* [1992] and Nydal *et al.* [1991]. Manolis [1995] presents “push in" experiments. The gas entrainment rate is obtained by a balance performed over the slug front:

$\dot{V}_{FR} = \frac{V_{FR}}{A} = k_1 W_i \left(\left(U_T - U_{Lfilm} \right) - \left(U_T - U_{Lfilm} \right)_{MIN} \right)$. Much more recently, Brauner *et al.* [2004] presented a new model for vertical flow to predict entrainment of bubbles from the Taylor bubble tail and their re-coalescence at the following Taylor bubble nose. This model is based on a balance between the turbulent kinetic energy and the rate of bubble surface energy production. Guet [2006] proposed a new model for the rate of gas entrained from the Taylor bubble to the liquid slug for vertical and inclined flows. This model uses the work done by the pressure force at the rear of the Taylor bubble. An iterative approach is then employed to equate the gas entrainment flux and the gas flux obtained *via* the conservation equation.

- **Gas-Liquid-Liquid Flow**

Work has been performed recently on this subject by Bonizzi *et al.* [2004]. Three different approaches are presented. The first is based on the two-fluid model, assuming homogeneous flow between the liquids. The other two are based on a drift-flux concept introducing local slip between oil and water. The drift flux method seems to fit the experimental data better.

1.3.4 Complex Phenomena

Changes in the geometry of the line have a direct impact on the behaviour of the main flow. For example, a converging or diverging junction will change the flow pattern. The phenomenon of "severe slugging", a non-stable production event, may also occur. This can pose serious problems, because large and fluctuating rates of gas and liquid can severely reduce production and in the worst cases shut down or damage topside equipment such as separator vessels and compressors.

Consequently, predicting slug characteristics is essential for the optimal, efficient, safe and economical design and operation of multiphase gas-liquid slug flow systems.

Models are needed to design current "Slug Catchers" (diameter, length) in order to control and dissipate the energy of the incoming gas stream, reducing fluid velocities to promote a "stratified" flow regime and subsequent gravity separation.

1.3.4.1 Gravity Induced Slug

The offshore seabed and onshore land are composed of horizontal, uphill and downhill sections. The production lines, following this land, consist of interconnections of different angles. Transient flow behaviour due to this geometry takes place in the pipe. The result is terrain-induced slugging which can cause severe operational problems. The conditions for the appearance of terrain/severe slugging are an accumulation of liquid at low points promoted by low flow rates and stratified flow in the upstream pipe leading down to the low point. The liquid slowly blocks the pipe section causing the upstream gas pocket pressure to increase critically, until its value is great enough to push the liquid slug as shown in Figure 1.17.

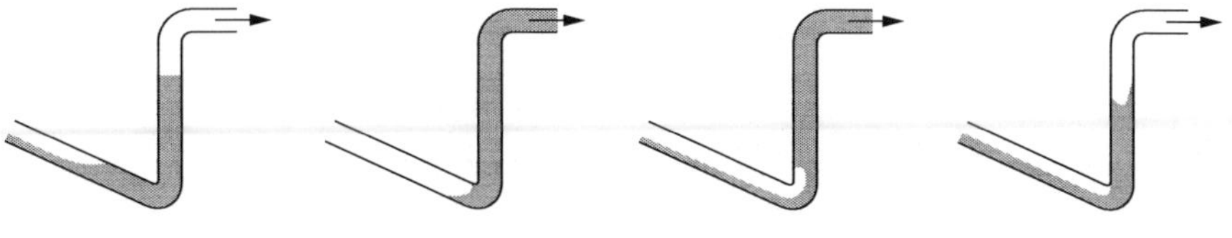

Figure 1.17

Severe slugging phases

The most severe form of terrain slugging occurs for a system pipeline-riser and is called "severe slugging". A rough rule predicts that for riser-induced slugging, the slug length equals one to three riser depths. Slug length models show that this rule is not valid for deep water (2,000 m). Schmidt [1977] was the first to describe severe slugging in four steps: slug formation, slug movement (production) into the separator, blowout, liquid fallback (Fig. 1.14). Boe [1981], Taitel [1986], Taitel *et al.* [1990] presented a stability analysis to predict the zone where this phenomenon existed. Vierkandt [1988] found Taitel *et al.* [1990]'s stability analysis to be inadequate. Hence extension of this analysis to terrain slugging is not expected to give accurate results. Serica [1990] proposed an alternative model based on his severe slugging model.

Other studies have been done on the control or elimination of the slugging, based on back-pressure increase (Yocum [1973]), gas-lift system (Hill [1990]), or choking. More recently, Jansen [1990] presented a comparison between the efficiency of various control systems.

1.3.4.2 Junction Flow

Due to the considerable development of production networks, the various of junction shapes (tees, Y…) have to be integrated into the modelling studies as they have an impact on the flow pattern and the total pressure drop. There are different kinds of junctions: converging or dividing junctions, symmetric or asymmetric junctions (see for example Fig. 1.18).

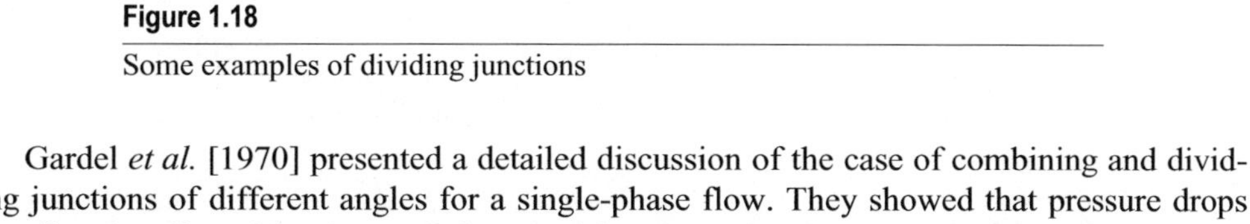

Figure 1.18

Some examples of dividing junctions

Gardel *et al.* [1970] presented a detailed discussion of the case of combining and dividing junctions of different angles for a single-phase flow. They showed that pressure drops are directly affected by the angle between the pipes, the diameter ratio and the degree of rounding of the corner.

Multiphase flow is far more complex to study as the pressure drop is heavily dependent on the flow pattern. Azzopardi [1999] makes an extensive review of dividing gas-liquid flows. Schmidt *et al.* [1994] present a review of the existing models in the literature.

1.3.5 Thermodynamic Modelling

Calculation of the thermodynamic properties of petroleum fluids is an essential step in the modelling of multiphase flow in pipelines. Thermodynamic modelling aims to determine the number of phases in equilibrium, as well as the phase properties, for given operating conditions (e.g. pressure, temperature and fluid composition). A reliable estimate of these properties is required for accurate hydrodynamic modelling.

The main characteristics of the thermodynamic behaviour of petroleum fluids will be presented in the first section of this chapter. The next sections will then present an overview of the various methods that exist for modelling of the thermodynamic behaviour. Three main features will be discussed: the thermodynamic model, a description of the mixture used as the model input and the numerical method used to solve equilibrium problems assuming a thermodynamic model and a fluid description. It should be noted that the discussion in this section assumes that phase equilibrium is reached at all times.

1.3.5.1 Fluid Behaviour

A typical Pressure-Temperature diagram for a reservoir fluid mixture is shown in Figure 1.19 (adapted from Riazi, [2005]). Depending on the pressure and temperature conditions, a typical fluid can comprise a single phase or two phases. The two-phase liquid-vapour region is located within the envelope (bold line) shown in Figure 1.19. The bubble point curve and the dewpoint curve merge at the critical point, where the liquid and vapour phases in equilibrium are identical. The dotted lines in Figure 1.19 indicate the amount (in vol. %) of the liquid phase.

A conventional petroleum fluid can be classified according to the position of the point representing Pressure-Temperature reservoir conditions with respect to the phase envelope (see Figure 1.19):

- Oils are characterized by a reservoir temperature located below the critical temperature of the mixture. In such cases, when the pressure of the liquid is reduced at constant temperature (from point A to point B in Figure 1.19), vaporisation begins at the bubble point pressure.
- Gas condensates are characterized by temperatures between the critical temperature and the cricondentherm temperature (highest temperature of the phase envelope, as shown in Figure 1.19). The first drop of liquid can appear by either reducing (from C to D) or increasing (from E to F) the gas pressure at constant temperature. Such mixtures exhibit two dew points and no bubble point.
- In the case of gases, reservoir temperatures are located above the cricondentherm temperature (point G) and the mixture cannot be liquefied when pressurised at constant temperature.

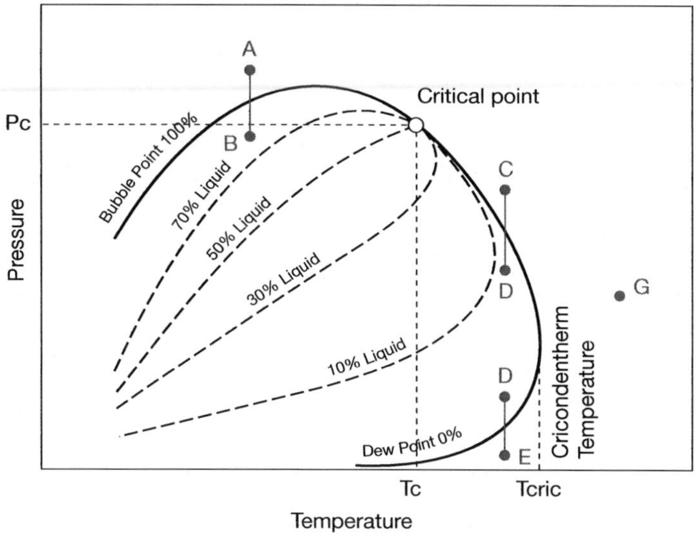

Figure 1.19

Typical PT diagram for a petroleum fluid mixture

In the case of fluids containing CO_2, the mixture can exhibit more complex behaviour with some possible liquid-liquid or liquid-liquid-vapour equilibrium at relatively low temperatures.

A typical phase diagram of a natural gas or oil mixed with water is shown in Figure 1.20 [Vidal, 2003]. As seen previously in the case of water-free mixtures, the liquid-vapour equilibrium curve of the hydrocarbon mixture (dotted line) is split into dew and bubble point curves that meet at the critical point. A three-phase region is now located inside this envelope, corresponding to the coexistence of either liquid water or hydrate, in addition to the classical liquid-vapour equilibrium. Hydrates are clathrate compounds formed by light hydrocarbon molecules trapped within a crystalline structure made up of water molecules (see Sloan [1998]). The solid line in the phase diagram shown in Figure 1.20 marks the Pressure-Temperature plane in two regions: a low-temperature, high-pressure domain where hydrates are thermodynamically stable and a low-pressure, high-temperature domain where hydrates cannot form. The presence of salts (electrolytes) or methanol in the water phase will modify the hydrate stability conditions, and all the properties of the aqueous phase.

There are other types of equilibrium involving petroleum fluids and a solid deposit: paraffin crystallisation and asphaltene flocculation. In all these cases, specific models capable of predicting conditions at which deposits occur are required [Ballard, 2002; Coutinho, 1995; Firoozabadi, 1999]. The formation of these solids inside wells or transport pipes can have disastrous consequences.

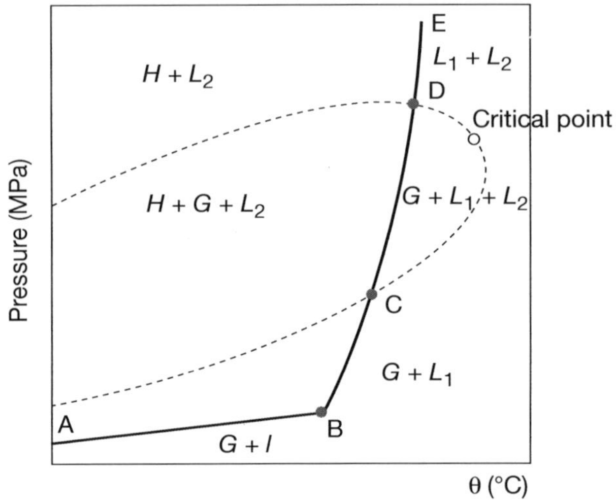

Figure 1.20

Phase diagram for a petroleum fluid and water mixture (from Vidal, 2003) G stands for gaseous hydrocarbon, L1 for liquid water, L2 for liquid hydrocarbon, H for hydrate, and I for ice

1.3.5.2 Thermodynamic Models for Fluid Property Calculation

Models are tools used to describe the physical behaviour of a material system. From purely empirical correlations to very complex equations based on statistical thermodynamics, a wide range of tools is now available in most multiphase flow simulators. The classification proposed in this section is taken from de Hemptinne *et al.* [2006]. Only the most commonly used models are described, the reader being invited to refer to adequate textbooks for more details [Vidal, 2003; Prausnitz *et al.*, 1999].

A thermodynamic model uses parameters to express a relationship between one or more state variables (e.g. pressure or temperature). It may undergo mathematical processing (differentiation, integration, etc.) before it actually delivers the property that is asked for. It is important to stress that adequate parameters must be available. This issue is particularly important for petroleum fluids and will be further discussed in the next section.

We can classify the models used according to their physical basis, as follows:

- correlations,
- activity coefficient models,
- equations of state.

- **Correlations**

A correlation is defined as a purely mathematical expression that enables a single property to be calculated on the basis of state variables and numerical parameters. Correlations are normally used for data interpolation (or, in the worst case, extrapolation, preferably not too far from the experimental domain). In some cases, correlation uses the corresponding state principle. This principle states that the residual thermodynamic (e.g. compressibility, etc.) or thermophysical (e.g. viscosity, heat conductivity, etc.) properties of any component can be known if its reduced pressure ($P_r = P/P_c$) and temperature ($T_r = P/T_c$), as well as the acentric factor (ω) are given. This is obviously an oversimplification, but it would seem that for hydrocarbon fluids this principle is a very good approximation.

Some examples of correlations are given in Table 1.2. They are grouped according to their domain of use (physical property calculation, equation of state parameter determination, mixing rules).

Table 1.2 Some examples of correlations

	Correlation	Nbr of numerical parameters	State variables required
	Pure component physical properties		
Vapour Pressure Poling *et al.* [2001]	Antoine	3	T
	Wagner	4	T
Liquid volume	Rackett [1970]	3	T
Ideal Gas Heat Capacity	Passut & Danner [1972]	4	T
	Aly & Lee [1981]	5	T
Henry constant	Harvey [1990]	3	T
Oil-Gas surface tension	Macleod [1923]	2	ρ^L, ρ^V
	Phase physical properties		
Oil phase thermal conductivity	Campbell [1984]	2	T
Oil and gas phase viscosity	Lohrentz, Bray & Clark [1964]	5	T, ρ, xi
	Equation of state parameters		
Alfa function in cubic eos	Soave (1972)	1	T
	Mathias & Copeman [1983]	3	T
	Twu [1995]	3	T
Volume translation in eos	Ungerer [1997]	4	T
	Mixing rules		
Linear mixing rules	Kay rule for pseudo-critical parameters [1936], For the covolumes in eos…	0	xi, yi
Quadratic mixing rules	For the attraction parameter in eos	1	xi, yi

- **Activity Coefficient Models**

When the components comprise significant non-ideality (as in water-alcohol-hydrocarbon mixtures), the excess Gibbs energy models (G^E models) often provide a good approach. In general, they are expressed as

$$\gamma_i = f(T, \mathrm{x}, \mathrm{p}) \quad (1.17)$$

where γ_i is the activity coefficient, T the temperature, x the composition vector of the liquid phase and p a set of parameters that depends on the components in question. These parameters, which are considered to represent the molecular interactions between dissimilar species, must be determined by fitting their values to experimental mixture data. When no experimental data are available, the UNIFAC [Fredenslund *et al.*, 1975, 1977] method, which is based on the groups contribution concept, can be used.

From expression (1.17), it is possible to calculate either the excess Gibbs energy (G^E) or the excess (or mixing) enthalpy (H^E). Note that because these models are independent of pressure, excess volume is considered to be zero. This is why these methods cannot be used, as such, for high-pressure applications (not above 1.5 MPa).

Depending on the approximations made when calculating the excess enthalpy and the excess entropy, these activity coefficient models are classified as regular solution, athermal solution, local composition, and so on. Table 1.3 gives some examples of such activity coefficient models. The application range (always for a condensed phase) and the number of empirical parameters are given. For typical transport applications, the models that are most often used are probably UNIQUAC and NRTL. This is because they are also appropriate for describing liquid-liquid equilibria and are well-suited to calculating excess enthalpies.

In the particular case of electrolyte mixtures (water + salts), activity coefficient models also are very useful. Either specific approaches have been developed: Pitzer, [1973], or classic NRTL models have been adapted: Austgen *et al.* [1989] or UNIQUAC: Nicolaisen [1993].

- **Equations of State**

These equations, expressed as:

$$\mathrm{P} = f(\mathrm{T,V,n,a,b\ldots}), \quad (1.18)$$

are not only used for calculation of volumetric properties but, through derivation or integration, also allow computation of other thermodynamic properties such as fugacity, heat capacity, etc. [Prausnitz *et al.*, 1999; Vidal, 2003].

Table 1.4 presents the most important equations of state currently used in the petroleum industry. It provides some insight into their advantages and drawbacks. The number of empirical parameters required for their use is highlighted, as it is an indication of their possible accuracy and/or predictive power.

For pure components, the more complex the equation, or the more parameters it contains, the better it behaves for the derivative properties (heat capacity, isothermal compressibility, isobaric thermal expansivity, etc.) within its domain of validity. This is the case for the equations that extend the Virial equation of state; in particular, the BWR (Benedict *et al.*

Table 1.3 Activity coefficient models and their application range*

Type of model	Nbr of binary parameters	Advantages/Application domain	Drawbacks
Empirical models: Margules Van Laar	 1 to 3 2	• Simple • Parameters available for many binaries	• No extrapolation capacity
Regular solutions: Scatchard-Hildebrand	0	• Predictive (no binary parameters)	• Non-polar molecules of similar size • Only positive deviations to ideality
Athermal solutions: Flory	0	• Predictive (no binary parameters) • Solutions of polymers	• No enthalpy of mixing
Polymer solutions: Flory-Huggins	1	• Solutions of polymers • Includes enthalpy of mixing	• Only for binary mixtures • Equilibrium data required
Local compositions: NRTL UNIQUAC Wilson	 3 to 8 2 to 4 2 to 4	• Complex mixtures, • Include LLE (except Wilson) • Also able to calculate H^E	• Equilibrium data required • Correlation among parameters possible
Electrolytes: Pitzer	2	• Many parameters available	• Purely empirical: data must be available

* *The corresponding equations, together with their references, can be found in the Poling* et al. *[2001] review book.*

[1940]) type equations of state that were made readily usable for hydrocarbon mixtures by Lee & Kesler [1975]. Soave [1995] proposes a corresponding states method for the BWR equation, thus extending its application range to all petroleum components and their mixtures.

However, present-day petroleum engineers who need to describe such derivative properties for complex mixtures often use a simple cubic equation of state [Van der Waals, 1873; Vidal, 2003]:

$$P = \frac{RT}{v-b} - \frac{a(T)}{\left(v - br_1\right)\left(v - br_2\right)} \tag{1.19}$$

The reason is the simplicity and the flexibility of these equations [Twu *et al.*, 2004]:

– The corresponding states principle allows calculation of the pure component parameters.

Table 1.4 Most important equations of state with their scope of application

Equation type	Nb of pure component parameters	Nb of binary interaction parameters	Advantages	Drawbacks
Virial [Tsonopoulos, 1974] [Hayden & O'Connell, 1975]	1	1	• Very simple	• Applies only to the vapour phase below 1.5 MPa
BWR [Benedict *et al.*, 1940] MBWR [Setzmann & Wagner, 1991]	16 32	–	• Very accurate	• Parameters only available for pure components • Their regression requires many data • Require binary mixture data
SBWR [Soave, 1995]	3	3	• Based on the corresponding states principle	
[Lee – Kesler – Plocker Plocker *et al.*, 1978]	3	1	• Based on the corresponding states principles	• Not adapted to phase equilibrium calculations, particularly in the near critical region
Cubic equations, [Vidal, 2003]: VdW, PR, SRK, SW, PRH with temperature dependence with volume translation	2 3 or more		• Very easy to handle • Based on the corresponding states principle	• Not very accurate for liquid volumes if used without volume translation
classical mixing rules	4 or more	1 (2 sets for SW)		
G^E – based mixing rules		2 to 6		• Require binary mixture data
CPA [Kontogeorgis *et al.*, 1996]	3 to 5	1	• For associating fluids (water, alcohols…)	• Pure component parameters not commonly available • Binary parameters to be regressed on mixture data
SAFT [Chapman *et al.*, 1990]	3 to 5	1	• Parameters may be calculated by contribution of groups • Well-suited to long-chain molecules (polymers, etc.)	• Complex • Seldom used so far
including polar terms	4 to 6		• May be used with polar molecules	

- For phase equilibrium calculations, the temperature dependence of the $a(T)$ parameter can be adjusted to any vapour pressure curve using one of the correlations shown in Table 1.2 (eos parameters section).
- If volume (or any single-phase property) predictions are needed, a volume translation parameter can be used [Peneloux *et al.*, 1982; Ungerer *et al.*, 1997].
- Moreover, extending equations of state to mixtures relies on the use of mixing rules for their parameters (a, b...). In Table 1.4, the mixing rules for the intermolecular attraction parameters $\left(a = \sum_i \sum_j x_i x_j \sqrt{a_i a_j}\left(1 - k_{ij}\right)\right)$ and the covolumes $\left(b = \sum_i x_i b_i\right)$ are called "classical mixing rules" and are well-suited to all non-polar mixtures and to hydrocarbon-acid gas mixtures. The binary interaction parameters k_{ij} (BIP) are best fitted to binary mixture data, but it is often convenient to use correlations published in the literature: Kordas *et al.* [1995] propose correlations for estimating the CH_4-heavy hydrocarbons BIP, Carroll & Mather [1995] for the H_2S-hydrocarbons BIP, and for CO_2-hydrocarbons, a single value of 0.13 seems to be generally accepted [Valderrama *et al.*, 1988]. If strong non-idealities exist in the system, for instance when alcohols (methanol or glycols) have been added to the reservoir mixture, a G^E based mixing rule must be used [Wong & Sandler, 1992; Michelsen, 1990; Huron & Vidal, 1979].

Even though the accuracy of these classical equations of state may be sufficient, their predictive power for complex and polar molecules remains limited. This is why more predictive approaches are now being developed, on the basis of statistical mechanical principles. For instance, the Cubic Plus Association equation of state from Kontogeorgis *et al.* [1996] combines the simplicity of a cubic equation with an additional contribution that is related to molecules forming associations. It remains a semi-empirical equation but allows a much more predictive representation of systems containing associating components (water, alcohols, etc.).

The problem becomes still more complex when ionic species are present in the aqueous phase. Liu & Watanasiri [1999] have published an interesting review of this problem. The Peng-Robinson [1976] cubic equation of state has been adapted by Soreide & Whitson [1992] to deal with the case of water-hydrocarbon mixtures with sodium chloride. If other ionic species are present, an activity coefficient model [Pitzer, 1995; Nicolaisen *et al.*, 1993] should be used.

Finally, the presence of large amounts of hydrogen requires a specific approach. [Moysan & Vidal, 1986] proposed BIP correlations for use with a cubic equation of state. This approach provides very good results for the solubility of hydrogen in petroleum cuts. However, for mixtures such as these, the generally accepted method in the refining industry is that of Grayson & Streed [1963].

1.3.5.3 Fluid Description

In this section, we will see how complex petroleum fluids can be described by a compositional representation used as an input to thermodynamic models. Representations such as these correspond to a given number of pure components and petroleum cuts, along with their characteristic parameters needed by the thermodynamic model (i.e. critical parameters, acentric factors, molecular weights, binary interaction parameters, and so on). We will also discuss how to reduce this number of components in order to avoid prohibitive calculation times during the multiphase flow simulation.

- **Detailed Compositional Description and Associated Characteristic Parameters**

The specificity of petroleum fluids results from the fact that they contain several hundreds of components, mainly hydrocarbons, from methane to complex molecules made up of more than 100 carbon atoms. The molecules composing the light end, usually up to C_6 or C_{10} (depending on the application), are all individually identified using gas chromatography and their characteristic properties are well known. Beyond this limit, i.e. usually from C_7 or C_{11}, only the major families (paraffinic, naphthenic and aromatic) are identified for each carbon atom number and the fluid is described in terms of sets of compounds called petroleum cuts.

When using compositional thermodynamic models, both pure component and petroleum cut parameters, as well as binary interaction parameters, are needed to compute the mixture properties. These parameters can originate from different sources, depending on the type of component considered.

- Pure component parameters can be found in databases. The characteristic parameters of the most common components can be found in the Poling *et al.* [2001] book or in electronic data banks like that of the DIPPR (Design Institute for Physical Properties) or the DETHERM (proposed by Dechema). However, all commercial simulators also provide parameter values so that the models proposed can be used with a large variety of components without any additional input from the user.
- Petroleum cut parameters can be estimated using correlations (Table 1.5) or group contribution methods [UNIFAC – Fredenslund *et al.*, 1975 – for G^E models; Coniglio *et al.*, 2000, for cubic eos] provided some information about the petroleum cut is available (typically boiling temperature and density for distillation fractions, or molecular formula for complex molecules). These parameters, especially in the case of heavy cuts, can be further adjusted in order to reproduce measured properties of the overall mixture originating from PVT studies (e.g. dew and bubble points, liquid viscosity, Gas Oil Ratio…). Here, we must stress the importance of the quality and quantity of the available data for reliable modelling.

Usually, this detailed representation of the fluid, based on several pure components and petroleum cuts, cannot be used directly in simulators because the calculation would be too time-consuming. Different approaches exist in order to reduce the number of components to be considered. In some cases, Black-Oil approaches are still employed, reducing the real mixture into a pseudo-binary system. Increasingly frequently, pseudo-compositional (also called lumped) descriptions are used, based on a limited number of pseudo-components.

Table 1.5 Most common methods for calculating characteristic parameters for petroleum cuts starting from their specific density and their mean boiling temperature (see nomenclature for properties definition). More details can be found in Riazi [2005]

	Riazi *et al.* [1998]	**Twu [1984]**	**Riazi-Daubert [1980]**	**Riazi-Daubert [1987]**	**Cavett [1962]**	**Kesler-Lee [1976]**	**Winn [1957]**	**Others**
T_c	x	x	x	x	x	x	X	
P_c	x	x	x	x	x	x	X	
ω						x		Edmister [1958]
Z_c	x	x	x	x				Rackett [1970]
MM		x	x	x		x		
$v^L(T)$					x			Rackett [1970]
$C_p^{\#}(T)$					x	x	X	
$P^{\sigma}(T)$					x			

- **Black-Oil Description**

Considering the large number of calculations needed in flow simulations, a very simple description can be used, based on a two-phase and two-component system (see Figure 1.21).

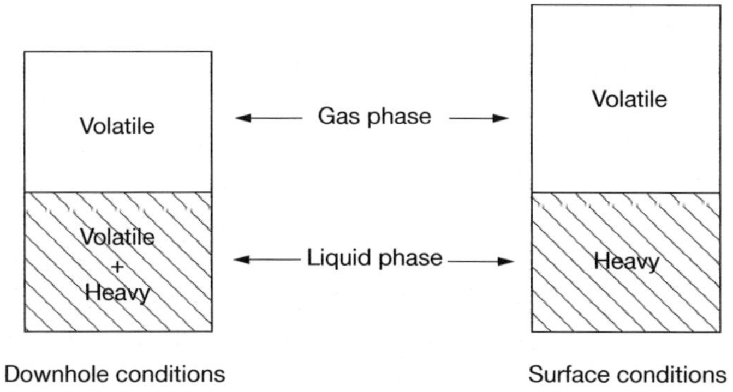

Figure 1.21

Schematic representation of the Black-Oil description

The two components involved in this description are a heavy component, which has the characteristics of stock tank oil, i.e. the oil as it is collected from the well, after all the gas has been removed at surface conditions, and a volatile component, whose properties are those of the surface gas. In the simplest Black-Oil version, the gas phase comprises the volatile component alone, whereas both components can be found in the liquid phase. Such a simple approach cannot be applied to volatile oil or gas condensate. The minimum information required for the Black-Oil description is the gas specific gravity, the stock tank oil specific gravity and the Gas Oil Ratio. The phase properties and phase quantities are then determined based on very simple PVT correlations (see Petroleum Engineering Handbook [1998] for a review of many of these correlations). When using such an approach, special attention must be paid to the validity domains of the correlations.

The Black-Oil description has the advantage of saving considerable amounts of computation time and can be used in cases where very little compositional data for the fluid is available.

- **Lumped Pseudo-Compositional Description**

Another method commonly used to describe petroleum fluids for thermodynamic modelling is to lump components into a small number of pseudo-components. It is essentially based on an appropriate fit of the thermodynamic behaviour to the detailed mixture. The procedure for describing a complex fluid by a set of pseudo-components is summarised by a two-step approach:

1. Firstly, if a fixed number of pseudo-components is required, the best arrangement for grouping components must be found. There are several lumping methods, ranging from simple clustering based on carbon atom numbers or boiling point temperatures to automatic aggregation methods such as the "dynamic clouds" method proposed by Montel [1984]. Whatever the method, the number of pseudo-components is always a compromise between the accuracy needed for the properties to be computed and the calculation time acceptable for the simulation.
2. Secondly, the pseudo-components must be given adequate characteristic parameters as input into the model. These parameters (e.g. critical coordinates, acentric factors, binary interaction parameters, etc.) can be determined by a number of correlations, from linear weighting [Kay, 1936] to more complex mixing rules [Lee-Kesler, 1975; Pedersen et al., 1984; Leibovici, 1993].

An example of use of this pseudo-compositional approach in a multiphase flow simulation where severe slugging occurs can be found in Henriot [1999]. In this example, a fluid description based on four pseudo-components was successfully used to assess the liquid accumulation phenomenon within a reasonable computing time.

The drawback of these simplified fluid representations (Black-Oil or pseudo-compositional representation) is the loss of the detailed compositional information necessary in many situations, such as in surface process engineering or in solid deposit prediction. A solution, called "delumping", has been proposed for restoring detailed compositional information from the results of a lumped compositional simulation [Leibovici, 1998] or even a Black-Oil simulation [Barroux, 2000]. Nevertheless, the most accurate way to describe the most difficult situations, especially when BIP are required, remains the use of a detailed composition.

1.3.5.4 Phase-Equilibrium Calculation

After discussing the choice of an appropriate thermodynamic model associated with an appropriate fluid description, the final concern in thermodynamic modelling is the numerical method to be used in order to solve the equilibrium problem. Equilibrium calculations performed in industrial simulators are based on an algorithm that uses the fact that the chemical potentials or fugacities of each component in all phases are equal:

$$f_i^L(T,P,x) = f_i^V(T,P,y), \qquad i = 1,\dots,n . \tag{1.20}$$

These $2n$ equilibrium equations, coupled with mass conservation equations:

$$n_i = n_i^L + n_i^V , \qquad i = 1,\dots,n , \tag{1.21}$$

are the basis for calculating two-phase equilibrium at constant pressure and temperature.

- **Homogeneous and Heterogeneous Approaches**

The fugacities required in equations (1.20) can be calculated through different models. Either the same model is used for all phases (homogeneous approach), or a specific model is used for each phase (heterogeneous approach). Table 1.6 summarises the two approaches in the case of vapour-liquid equilibrium.

Table 1.6 The heterogeneous and the homogeneous approaches for liquid-vapour equilibrium calculation

	Homogeneous approach	**Heterogeneous approach**
	$\left(\varphi_i^L, \varphi_i^V\right)$	$\left(\gamma_i^L, \varphi_i^V\right)$
Advantages	High pressures. Phase envelope calculation, including the near-critical region.	Improved accuracy because of the large choice of models.
Drawbacks	Limited by the choice of an appropriate mixing rule.	Limited to pressures below 1.5 MPa. No phase envelope calculation.

The homogenous approach can only be applied with equations of state and enables the entire phase envelope to be calculated. With the advent of new equations of state as well as of improved mixing rules, there is a tendency towards greater use of this approach. However, in the case of strongly non-ideal mixtures (e.g. presence of components like water, salts, or alcohols) or in the case of mixtures involving a solid phase, it is common practice to use different models for the different phases (e.g. Soreide & Whitson for hydrocarbon-water systems).

- **Solution Methods**

Various methods are available for solving equations (1.20) and (1.21). The most commonly used methods will be briefly listed here for a liquid-vapour equilibrium calculation at constant pressure and temperature. For more details, the reader is referred to the books of

Firoozabadi [1999] and Michelsen & Mollerup [2004]. All these methods assume that the number of phases in equilibrium is known in advance.

Successive substitution technique.

This is an iterative procedure in which the unknowns are the phase molar fractions. Each iteration step consists in the calculation of these phase molar fractions by solving the Rachford-Rice [1952] equation, followed by an update of the equilibrium coefficients $K_i\left(K_i = y_i/x_i\right)$. The Wilson [1969] correlation is often used to guess the initial values of K_i.

Newton method.

The nonlinear system of equations (1.20) and (1.21) can also be solved by the Newton method. This method is based on the Taylor series expansion of the above-mentioned equations. The required Jacobian matrix is made of the partial derivatives of fugacities with respect to composition. The Newton method is characterised by fast convergence speed but it requires good initial estimates. A few successive substitution iterations can provide an appropriate initialization.

Gibbs energy minimization.

Alternatively, equilibrium can be formulated as a Gibbs energy minimization problem. Indeed, the equality of fugacities (equations 1.20), coupled with material balance (equations 1.21), is a necessary condition for solving equilibrium problems, but not in itself sufficient. The stability of the system will be ensured only if the system of predicted phases has the lowest possible Gibbs energy.

- **Stability Analysis**

The solution methods listed above, assuming a fixed number of coexisting phases, are often successfully used in solving equilibrium problems. However, in the vicinity of the phase envelope, especially in the case of complex mixtures in which the number of possible phases is not previously known, such approaches can fail, leading to a local minimum of the Gibbs energy surface. In such difficult situations, a stability analysis is required. A stability analysis of this type, known as the Gibbs tangent plane criterion [Baker, 1981; Michelsen, 1982], is a numerical test for deciding whether a phase of a given composition is thermodynamically stable in given pressure and temperature conditions. This analysis is suggested as a preliminary step in equilibrium calculations of complex mixtures.

1.3.5.5 Conclusion

In this section, some general guidelines for thermodynamic modelling of petroleum fluids are given. However, due to the large number of methods available and the variety of industrial situations that could be encountered, an exhaustive review could not be presented. We focused on describing the most commonly used methods, and discussing their application range, accuracy, strength, limitations and implementation. It must be pointed out that the final result of thermodynamic modelling will depend to a large extent on the quality of the fluid description and the accuracy of the characteristic parameters required.

1.3.6 Thermal Aspects

The thermal behaviour of effluent is one of the main issues to be managed when designing an offshore field development. Risks concern the hydrate and wax deposits inside the pipes that may reduce the flow rate or even plug the pipes in extreme conditions. Risks also concern the operating conditions of the topside equipment which is not designed to withstand a wide range of operating conditions. This thermal behaviour and the associated risks depend on the fluid characteristics, the environmental conditions and the pipe characteristics. It is now clear that, as compared with conventional shallow water offshore, the construction and exploitation of deep sea and ultra deep sea offshore fields involve new and specific concerns,. One of the most critical problems is to guarantee flow at such water depths, for both steady state production and shutdown/restart conditions.

The first characteristic of deep sea flow lines is their need to be *efficiently thermally insulated*, in order to preserve an adequate production temperature on the FPSO topsides, and also to avoid solid deposits on the pipe wall: deep offshore fields for which profitable exploitation is forecast are often spread over very large areas, implying long flow lines along the sea bed, into low-temperature sea water. Moreover, due to the depressurisation inside extremely deep offshore risers, gas phase expansion and light cuts vaporisation produces severe cooling of the effluent produced.

Another concern specific to deep sea flow lines is the risk of gas hydrate plugs forming relatively high temperature under high pressure, especially during production shutdown and restart episodes. *The cool-down time* needed to reach the hydrate formation temperature in shutdown conditions is a major requirement of deep sea flow lines and must be long enough to allow the completion of flow line emptying operations, or start-up of dedicated heating.

Thermal insulation of offshore flow lines often remains the best way of keeping crude in flow conditions, or of avoiding hydrate formation or wax deposit for as long as possible. In many cases, deepwater fields involve all of these problems, that are particularly critical in production shutdown situations.

In order to meet these requirements, existing or emerging technical solutions can be placed in the following two categories:

- Thermal insulation: Pipe-in-pipe technology or coating technologies are used, depending on the required thermal specification. Due to the vacuum in the annulus between the pipes, pipe-in-pipe technology is thermally efficient but the weight of the pipe-in-pipe structure is sometime a limitation for deep water development. Blocks of syntactic foams can be applied around the pipes to perform both the buoyancy and thermal insulation functions. Different structures of syntactic foams are possible using pressure-resistant glass bubbles and soft or rigid matrixes leading to a wide range of pressure resistance. Expanded foams are not reinforced by glass bubbles and are often used in shallow water. Depending on the thermal performances of the materials, both the insulation function and the cool-down time function can be guaranteed. Phase-changing materials can also be used to enhance the cool-down time capability of an insulated system in both pipe-in-pipe systems or externally coated pipes or bundles.

- Passive thermal insulation + active heating: In this case, the two requirements, insulation and cool-down time, are dealt with independently. Active heating can be electrical or by fluid pumping from a thermal heater located on the topside. Electrical heating of the pipe also requires an energy source and good electrical connections. Thermal insulation is only required during the active phases of production while electric heating generally avoids effluent cool-down during the non-producing phases. Both flexible and rigid pipes are compatible with such combined systems.
- Additives are another means of avoiding hydrate or wax deposits. Low dosage additives can be used to reduce the weight of the insulation system in cases where the risk of deposit is limited to short period of time. Addition of large quantities of methanol displaces the thermodynamic domain of hydrate formation and keeps the effluent produced withinin a safe range. This type of technique entails certain environmental impact difficulties and the trend seems to be towards the use of anti-agglomerate additives allowing hydrates transportation in dispersed phase reducing the risk of hydrate plugs in the pipes.

In all cases, control of the thermal behaviour of the effluent is an essential issue and this chapter will present the main rules used to determine the fluid temperature profile.

1.3.6.1 Introduction

Let us consider a system with a multiphase fluid flowing in a pipeline made up of layers of various materials each with its own characteristics.

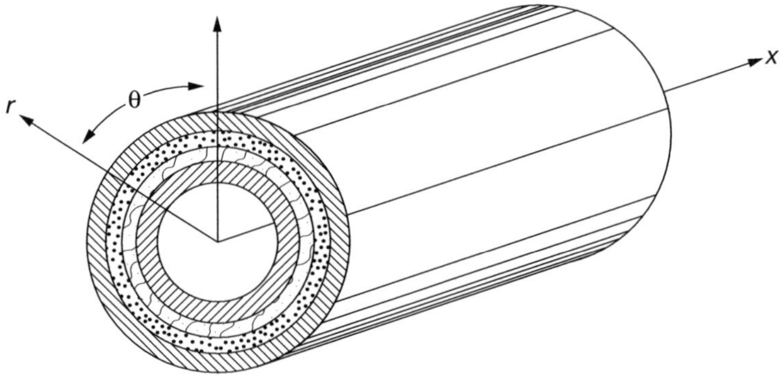

Figure 1.22

Pipeline wall sketch

Suppose that this composite cylindrical wall is composed of J materials of thickness $r_{j+1/2} - r_{j-1/2}$ and thermal conductivity λ_j.

The objective of this section is to determine the heat transfer between the multiphase fluid and the surrounding medium through the wall for a given abscissa x. It is characterised by a heat flux per unit length denoted Q_{in}^{W}. This flux is used in the right hand side of the energy balance to determine the transient fluid temperature profile along the pipe. Formally, it can be expressed as:

$$Q_{in}^{W} = 2\pi \left[\lambda(r)\, r \frac{\partial T}{\partial r} \right]_{r=r_{1/2}} .$$

Mainly, it depends on the fluid heat transfer coefficient, the surrounding medium behaviour and the pipeline wall thermal resistance.

1.3.6.2 Fluid Heat Transfer Coefficient

Let us consider the mixture flowing through the pipeline and assume that the inner surface is cooler than the fluid so that heat is being transferred from the fluid to the solid. The rate of heat flow across the solid-fluid interface is expected to depend on the area of the interface and the temperature drop between solid and fluid. Hence, we write the heat flux per unit length:

$$Q_{in}^{W} = 2\pi r_{1/2}\, h_{fluid} \left(T_{1/2} - T_{fluid} \right),$$

where $T_{1/2}$ is the inner wall temperature, T_{fluid} is the fluid temperature and the proportionality factor h_{fluid} is the heat transfer coefficient.

This coefficient h_{fluid} can be expressed as a function of the flow pattern, the thermal conductivity λ_k of each phase k of the mixture and the dimensionless Nusselt number Nu. For example, for gas-liquid flow, it reads:

$$\begin{cases} h_{fluid}^{S} = \dfrac{1}{2 r_{1/2}} \Big[p_{wG}\, \lambda_G\, Nu\left(\mathrm{Re}_G, \mathrm{Pr}_G, Gr_G\right) & \text{(for stratified flow)} \\ \qquad + p_{wL}\, \lambda_L\, Nu\left(\mathrm{Re}_L, \mathrm{Pr}_L, Gr_L\right) \Big] & \\ h_{fluid}^{D} = \dfrac{\lambda_L}{2 r_{1/2}} Nu\left(\mathrm{Re}_M, \mathrm{Pr}_M, Gr_M\right) & \text{(for dispersed flow)} \\ h_{fluid}^{I} = (1-\beta)\, h_{fluid}^{D} + \beta\, h_{fluid}^{S} & \text{(for intermittent flow)} \end{cases}$$

where Re_k, Pr_k and Gr_k are the Reynolds, Prandtl and Grashof numbers of the phase, k, p_{wk} is the dimensionless wall perimeter in contact with the phase k and β is the void fraction of the intermittent flow.

Moreover, this Nusselt number depends on the type of phenomena to be taken into account (forced convection, free convection, etc.). For instance, in case of forced convection, it reads:

$$\begin{cases} Nu = 4 & \text{(for laminar flow)} \\ Nu = 0.023\,\mathrm{Re}^{0.8}\,\mathrm{Pr}^{1/3} & \text{(for turbulent flow)} \end{cases}.$$

1.3.6.3 Surrounding Medium Behaviour

The surrounding medium is the heat transfer engine and its behaviour naturally depends on its nature. Indeed, the pipeline can be subject to temperature fluctuations (mountain, desert), can be in contact with a cool environment (sea bed) and can be partially or completely buried. To accurately take account of all these thermal phenomena generally requires linkage with a simulation software solving the heat equation with a mesh and numerical method. But most effects can be approximated by an outside heat transfer coefficient:

$$Q_{out}^{W} = 2\pi r_{J+1/2}\, h_{out}\left(T_{out} - T_{J+1/2}\right).$$

1.3.6.4 Overall Heat Transfer Coefficient

Insulation is generally used to limit heat transfer between the flowing mixture and the outside. Basically, the pipeline wall can be based on a foam layer surrounded by steel material. But more sophisticated methods can be used, such as electric heating, pipe-in-pipe or bundle system with secondary heating lines.

For a composite wall and if thermal effects have little or no time-dependency, it is possible to limit the problem to steady-state. Thus, the heat flux through each layer is the same and Q_{out}^{W} is equal to Q_{in}^{W}. In addition, the one-dimensional form of Fourier's law for heat conduction implies that:

$$\frac{\partial}{\partial r}\left[\lambda(r)\, r\,\frac{\partial T}{\partial r}\right] = 0\,.$$

This equation being time-independent, it does not take into account thermal accumulation inside the pipeline wall. Integrating it gives a constant that can be directly expressed as $\frac{Q_{in}^{W}}{2\pi}$:

$$\lambda(r)\, r\,\frac{\partial T}{\partial r} = \frac{Q_{in}^{W}}{2\pi}.$$

This equation can then be rewritten as:

$$\frac{\partial T}{\partial \ln r} = \frac{Q_{in}^W}{2\pi} \frac{1}{\lambda(r)}.$$

Let us now define $T_{j-\frac{1}{2}}$ the value of the wall temperature at $r = r_{j-\frac{1}{2}}$ and assume we are inside the jth layer. $r \in \left[r_{j-\frac{1}{2}}, r_{j+\frac{1}{2}}\right]$ implies that the thermal conductivity $\lambda(r)$ has a constant value λ_j. Integrating the last equation between $r_{j-\frac{1}{2}}$ and r gives:

$$T(r) - T_{j-\frac{1}{2}} = \frac{Q_{in}^W}{2\pi\lambda_j} \ln\left(\frac{r}{r_{j-\frac{1}{2}}}\right) \qquad \forall\, r \in \left[r_{j-\frac{1}{2}}, r_{j+\frac{1}{2}}\right].$$

Thus, the following figure shows the temperature profile inside the pipeline wall structure.

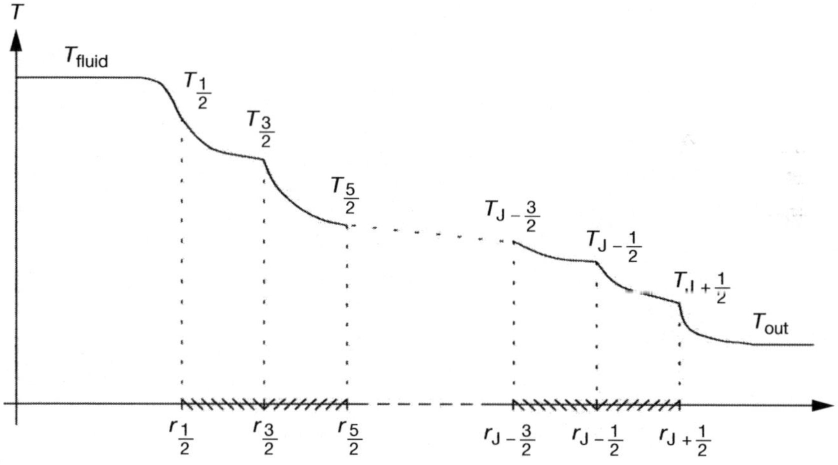

Figure 1.23
Wall temperature profile

Moreover, writing this equation for $r = r_{j+\frac{1}{2}}$ and assuming that $Q_{out}^W = Q_{in}^W$ leads to a system consisting of $J + 2$ algebraic equations:

$$\begin{cases} T_{\frac{1}{2}} - T_{fluid} = \dfrac{Q_{in}^{W}}{2\pi r_{\frac{1}{2}}} \times \dfrac{1}{h_{fluid}} \\[2ex] T_{\frac{3}{2}} - T_{\frac{1}{2}} = \dfrac{Q_{in}^{W}}{2\pi r_{\frac{1}{2}}} \times \dfrac{r_{\frac{1}{2}} \ln\left(r_{\frac{3}{2}} \big/ r_{\frac{1}{2}} \right)}{\lambda_1} \\ \vdots \\ T_{j+\frac{1}{2}} - T_{j-\frac{1}{2}} = \dfrac{Q_{in}^{W}}{2\pi r_{\frac{1}{2}}} \times \dfrac{r_{\frac{1}{2}} \ln\left(r_{j+\frac{1}{2}} \big/ r_{j-\frac{1}{2}} \right)}{\lambda_j} \\ \vdots \\ T_{J+\frac{1}{2}} - T_{J-\frac{1}{2}} = \dfrac{Q_{in}^{W}}{2\pi r_{\frac{1}{2}}} \times \dfrac{r_{\frac{1}{2}} \ln\left(r_{J+\frac{1}{2}} \big/ r_{J-\frac{1}{2}} \right)}{\lambda_J} \\[2ex] T_{out} - T_{J+\frac{1}{2}} = \dfrac{Q_{in}^{W}}{2\pi r_{\frac{1}{2}}} \times \dfrac{r_{\frac{1}{2}}}{r_{J+\frac{1}{2}}} \dfrac{1}{h_{out}} \end{cases}.$$

Addition of all equations gives an expression for $T_{out} - T_{fluid}$, which may be solved for Q_{in}^{W} to give:

$$Q_{in}^{W} = \frac{2\pi r_{\frac{1}{2}} \left(T_{out} - T_{fluid} \right)}{\dfrac{1}{h_{fluid}} + \displaystyle\sum_{j=1}^{J} \dfrac{r_{\frac{1}{2}} \ln\left(r_{j+\frac{1}{2}} \big/ r_{j-\frac{1}{2}} \right)}{\lambda_j} + \dfrac{r_{\frac{1}{2}}}{r_{J+\frac{1}{2}}} \dfrac{1}{h_{out}}}.$$

Let us now note $R_{fluid} = \frac{1}{h_{fluid}}$, $R_{wall} = \sum_{j=1}^{J} \frac{r_{\frac{1}{2}} \ln\left(r_{j+\frac{1}{2}} \big/ r_{j-\frac{1}{2}} \right)}{\lambda_j}$ and $R_{out} = \frac{r_{\frac{1}{2}}}{r_{J+\frac{1}{2}}} \frac{1}{h_{out}}$.

The heat flux reads:

$$Q_{in}^{W} = \frac{2\pi r_{\frac{1}{2}} \left(T_{out} - T_{fluid} \right)}{R_{fluid} + R_{wall} + R_{out}}.$$

Finally, this result can be rewritten in a form of Newton's law of cooling:

$$Q_{in}^{W} = 2\pi r_{\frac{1}{2}} U \left(T_{out} - T_{fluid} \right),$$

where the quantity U called the overall heat transfer coefficient (based on the inner pipeline surface) is given by:

$$U = \frac{1}{R_{fluid} + R_{wall} + R_{out}}.$$

1.3.6.5 Conclusion

The overall heat transfer coefficient is a good characteristic of pipeline insulation and can be used to determine the fluid temperature profile. But this approach does not take account of the thermal inertia of the wall, which is obviously predominant in transient cases (cool-down time computation for instance). Instead, the transient heat conduction equation has to be solved and this can mainly be done with a numerical solver.

In the case of a pipeline with at least one insulation layer, thermal resistances R_{fluid} and R_{out} can clearly be ignored. Thus, the inner wall temperature $T_{\frac{1}{2}}$ and the fluid temperature T_{fluid} are supposed to match exactly, as do the outer wall temperature $T_{J+\frac{1}{2}}$ and the outside temperature T_{out}. Furthermore, the U value can be directly approximated by:

$$U = \left(R_{wall} \right)^{-1}.$$

One final comment needs to be made concerning this development. We assumed the layers to be closely mated with no existing air spaces. It is clear that if the layers only touch each other at a few points, the thermal resistance will be significantly increased due to the high thermal resistance of these air voids.

For more detailed information on thermal calculations, please refer to "Heat Transfer" by Jacob or "Transport Phenomena" by Bird, Stewart and Lightfoot.

Chapter 2

Multiphase Pumping

2.1 INTRODUCTION

Initial work aiming at developing multiphase pumping in the oil industry was begun by IFP and Total in the mid-seventies to extend the application of down-hole electric submersible pumps (ESP) to gassy wells [Arnaudeau, 1982]. At the end of the development process, a multiphase pump prototype was manufactured by Pompes Guinard and tested by Total in an Indonesian well (1985).

In the eighties, interest was rekindled in multiphase pumps (MPP) with a view to reducing offshore development costs, especially in the North Sea. A number of development projects were launched in this decade as multiphase boosting appeared to be a key technology in increasing the tie-back distance of subsea satellite fields.

In 1983, BP, Mobil and Shell initiated the development of a twin-screw multiphase pump with Stothert & Pitt Fluid Engineering Ltd. A prototype was designed, manufactured and tested on a UK field [Dolan *et al.*, 1986]. This expertise was continued by Multiphase Systems PLC and led to the commissioning of a first twin-screw pump on an offshore platform in Malaysia for Shell (1989) and then to a second one in the North Sea for BP.

The Poseidon Project, launched by Total, Statoil and IFP in 1984 represented a major step in MPP development. This project led to development and testing in Tunisia of the helico-axial concept now called the Poseidon pump [Engelman and Torp, 1990]. It also contributed to testing of a twin-screw pump. The participation of Shell in helico-axial development must also be mentioned since it led to the first subsea MPP installed in the North Sea in 1993 [Darrigan *et al.*, 1995].

Many other significant achievements could be mentioned. Progress was often difficult, but the developers were determined and persistent enough to ensure ultimate success.

The number of installed pumps has considerably increased as the industry gained confidence with the accumulation of field experience, onshore, offshore and subsea. In 2004, the total number of multiphase pumps installed in oil fields reached some 500 units. They are deployed all around the world. Their size has also considerably increased and a single machine can have a capacity of 3,600 m^3/h.

This chapter presents the fundamentals of multiphase boosting: the main pump concepts, their operating principles, their respective advantages and drawbacks, and how they can be introduced in a production system. For this purpose, few application cases are described in the last section of the chapter.

2.2 OVERVIEW OF MULTIPHASE PUMPING

2.2.1 Benefits and Typical Applications

Multiphase pumping essentially consists in adding hydraulic energy to an unprocessed production stream. It makes greater pressure drops possible before separation in a processing facility. Greater pressure drops mean longer tie-back distances, as indicated above, with the possibility of reusing the existing facilities in a depleting field without a local offshore platform or floating facility.

This characteristic is of special interest in deep waters, as well as onshore, in remote or hostile environments. Giant or large deepwater fields can be developed with stand-alone production schemes, using a dedicated floating facility. However, the economical development of smaller fields greatly depends on the availability of production tie-in to an existing facility and thus of boosting the well stream to sustain high flowrate over relatively long distances.

The ability to boost multiphase fluids implies a number of other capabilities:

- increased flowrate in an existing export pipe-line;
- production of low-pressure wells in a common high-pressure process;
- reduction of well-head flow pressure to increase production or ultimate well production;
- production and marketing of associated gas formerly flared;
- production of reserves from economically marginal fields.

Even without the elimination of an offshore platform, multiphase pumping allows significant cost savings (at least 30%) through the simplification of conventional production facilities and a reduced footprint, both onshore and offshore [Dal Porto, 1996; Leporcher, 2001].

2.2.2 Types of Pumps

Many types of pumps are currently used in single-phase flow (Figure 2.1). They are usually split into three categories on the basis of their physical working principle:

- positive displacement pumps, also called volumetric pumps, which transfer a finite volume of fluid from a low-pressure side to a high-pressure side. This category can be subdivided into several types according to the principle of the transfer system: progressing cavity, rotary (with single or twin-screw), reciprocating (or piston pump), diaphragm pumps, vane pumps;
- rotodynamic pumps, also called centrifugal pumps, dynamic pumps or turbomachines. They transfer kinetic energy to the fluid in a rotating part (called the impeller) and transform this energy into potential energy in a static part. This category can also be divided into several sub-categories: axial, radial, mixed-flow, depending on the overall shape of the impeller. A centrifugal pump can be composed of a single impeller or be multi-staged with several impellers assembled on a common shaft and separated by fixed intermediate diffusers.

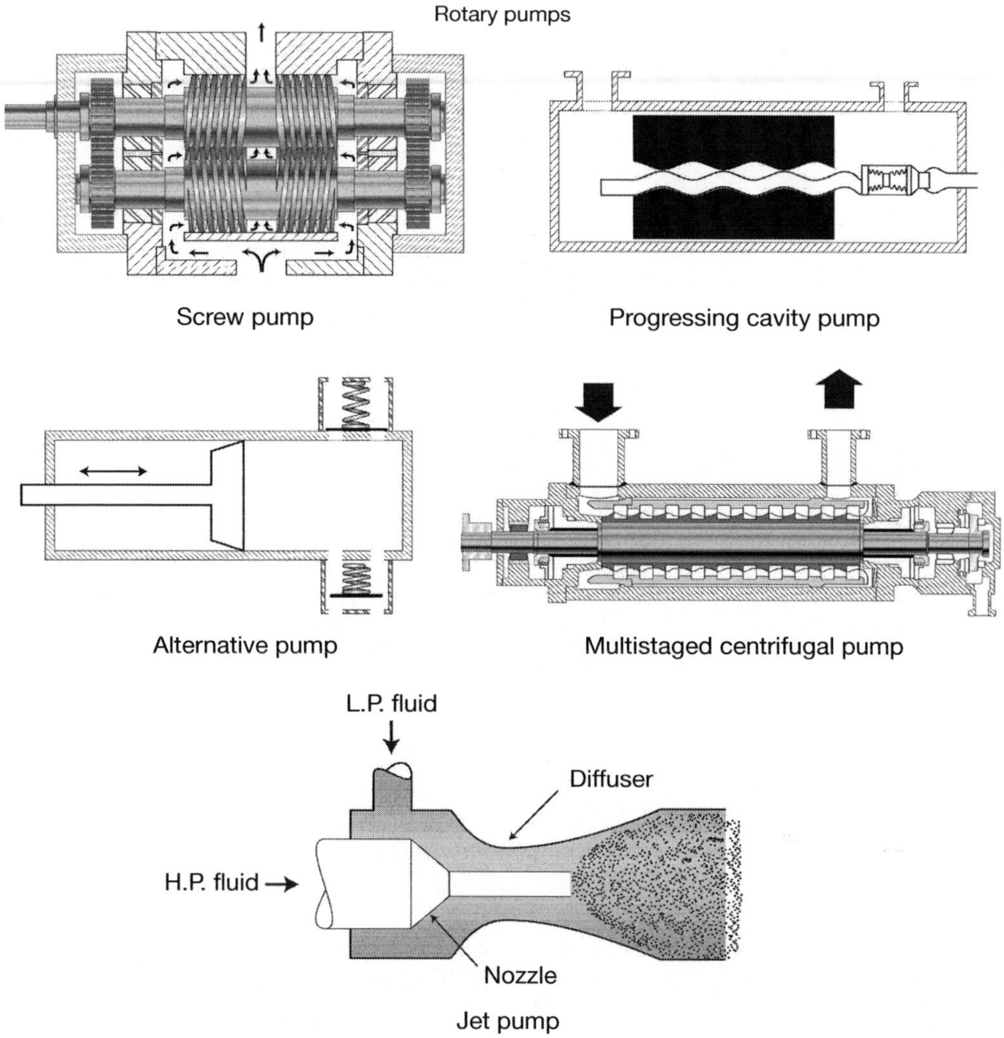

Figure 2.1

Some examples of pump types

- hydraulic pumps also called jet pumps or ejectors: a high-velocity fluid jet transfers its kinetic energy to a low-pressure fluid by mixing, then the mixture is decelerated in a diffuser nozzle to recover pressure. The driving and boosted fluids must be compatible.

Many concepts of every type have been proposed for multiphase flow operation. Many prototypes have been designed, manufactured and tested with varying degrees of success.

Three main concepts of multiphase pumps have emerged: the helico-axial concept that belongs to the rotodynamic pump category, twin-screw pumps and progressing cavity pumps that both belong to the positive displacement category. These three concepts are described and their principles explained in this chapter. Together, they today represent the largest number of MPP applications.

A few other types of MPP are also used for specific applications, such as reciprocating pumps and jet pumps. Reciprocating pumps (or piston pumps) can deliver high pressures but have a relatively low capacity. Jet pumps are used to boost low-pressure wells by high-pressure wells and combine their production [Sarsha, 2000; Andreussi *et al.*, 2003].

The respective advantages and drawbacks of positive displacement pumps and helico-axial pumps are outlined below. For multiphase applications however they complement each other, with each concept having its preferred areas of application:

- small capacity for progressive cavity pumps, as opposed to medium capacity for twin-screw pumps and large capacity for rotodynamic pumps;
- high viscosity liquid phase for progressing cavity and twin-screw pumps;
- down-hole or subsea applications for rotodynamic pumps;
- low suction pressure for volumetric pumps (twin-screws or progressing cavity).

These areas are in fact also those recommended for single-phase applications.

2.2.3 Main Issues of Multiphase Boosting

The boosting of unprocessed crude oil presents specific issues which are related to different features of unprocessed crude oil production:

- variation of flow conditions and operating domain on several time scales;
- the properties of gas, such as compressibility and thermodynamic behaviour, must be considered; MPPs are in fact hybrid machines, somewhere between compressors and pumps;
- fluid composition: the possible presence of solid particles (sand production, scale formation) generating abrasion and wear, the presence and concentrations of corrosive and toxic components (H2S), possible wax deposition.

2.2.3.1 Variation of Flow Conditions

A pump is often designed to operate at a given duty point and is normally selected so that its best operating conditions or efficiency are at or close to the duty point.

In oil production, it is often difficult to specify a single well-defined duty point. Flowrates, gas fractions, water cut, pressures, etc. follow short term variations as well as a long term evolution over the life of the field owing to reservoir depletion. Moreover, actual operating conditions can be different from initial forecasts due to various uncertainties and operating changes. All these flow variations and uncertainties broaden the range of possible operating points. This is why an MPP must offer a broad operating envelope. For oil production, flexibility is preferable to very good efficiency optimised in a narrow operating band.

A. Long-Term Operating Domain

Over the years, the production of a field evolves. A typical profile is shown on Figure 2.2.

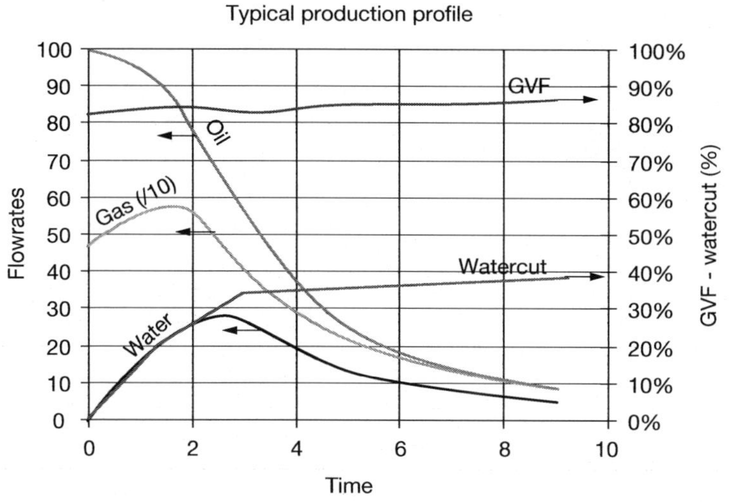

Figure 2.2

Typical long-term production profile

The production profile corresponds to a set of operating points which defines an operating domain for the pump. An MPP must generally accommodate a wide operating domain of flowrates and GVF.

Production predictions are generally based on extrapolation from well tests and reservoir studies during the exploration and delineation phase. For this reason, flowrate and pressure predictions often contain a degree of uncertainty and actual production data may be somewhat different.

To be able to implement the needed operating envelope, a variable-speed drive may be required, and sometimes a change of the pump internals may be planned after a certain period. Retrofitting can be more economical than choosing an initial operating domain that is too large. It can be also required if actual production is far from the initial forecasts and exceeds the adjustment capability.

B. Short-Term Flow Fluctuations and Slugs

Medium or short-term fluctuations of the operating point can be also observed under transient operations such as well start-up or shut-down, well tests, pipeline pigging operations, etc. Flow fluctuations on such time scales are also generated by hydraulic instabilities in wells or pipelines (hydrodynamic slugs, terrain-induced slugging, severe slugging in risers, etc.).

Figure 2.3 for instance shows a short-term record of total flowrate and pressure for a relatively stable well. Some wells may have unstable behaviour, as shown in Figure 2.4, where the well has cyclic behaviour characterised by relatively stable production periods alternating with a sequence of large gas pockets and liquid slugs. This is often observed with low-pressure wells in mature fields. The flow variations and uncertainties broaden the range of possible operating points for the pump.

In various situations, an MPP can be exposed to long liquid slugs or gas pockets. The pump group must be able to sustain high torque and high power demand in the event of liquid slugs and be tolerant to dry-running or low pressure rise in the event of gas pockets.

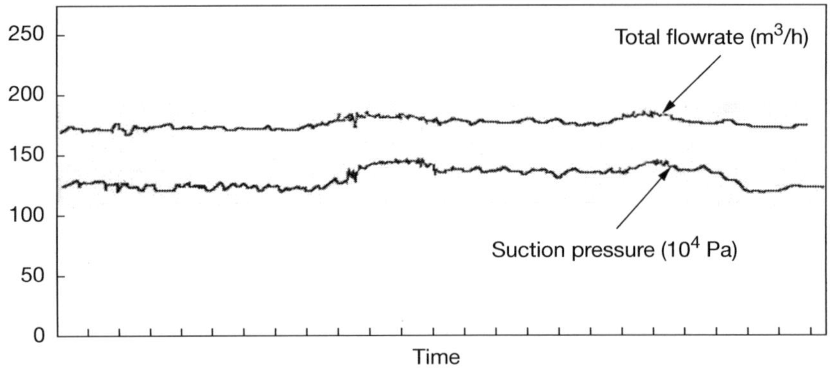

Figure 2.3

Example of inlet flow variations

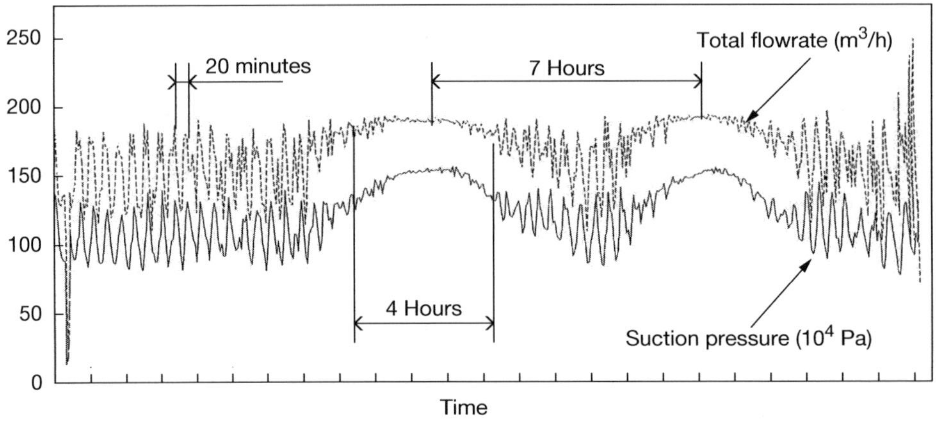

Figure 2.4

Other example of inlet flow variations

It is hard to predict the length and true density of slugs. They are neither fully liquid nor fully gas. Even though some pumps can, from a mechanical point of view, run continuously from 100% gas to 100% liquid, the power required is very different for the two extreme cases. It would not be economical to size the driver for the full power at 100% liquid. The driver would be oversized and would work most of the time at partial load.

For this reason, short-term flow variations are generally managed by a "process" approach, using a flow homogeniser or liquid recycle, described below in the "process control" section.

2.2.3.2 Gas Compressibility

The total flowrate capacity of the pump which is mentioned in the manufacturer's documentation is the total volumetric flowrate, the sum of actual liquid and gas volumetric flowrates in suction conditions. Depending on the suction pressure, the same pump can be used with very different gas mass flowrates or gas volumetric flowrates expressed in standard conditions.

The main visible effects of gas compressibility are the reduction of volumetric flowrate and the reduction of gas volumetric fraction (GVF) between the MPP inlet and outlet. Accordingly, the apparent mixture density increases. Depending on the pump type (positive displacement or helico-axial), the impact of these effects is somewhat different. They are described hereafter in the relevant section.

The thermodynamic properties of gas also have a number of important consequences, independent of the pump type. For instance, gas compression produces a temperature rise in the mixture. Usually the temperature rise in an MPP is low or moderate because the liquid cools the gas sufficiently. However significant temperature rises can be observed with a very high GVF mixture and high compression ratios, or in cases of low flowrate. Gas compressibility also influences the required compression energy of the gas phase. As with compressors, the relevant parameter which governs these effects is the compression ratio, *i.e.* the ratio of the discharge pressure to the suction pressure, and not the pressure rise.

2.2.3.3 Gas Re-Dissolution

The thermodynamic properties of the liquid and gas can also lead to mass transfer between the phases during compression. Mass transfer can be immediate or delayed. It is produced by re-dissolution of the gas in the liquid or condensation of the gas.

The re-dissolution of gas in oil influences the reduction in the gas volumetric flowrate and the reduction of gas fraction throughout the pump. Re-dissolution increases the effect of compressibility. It has a favourable effect in helico-axial pumps.

How much gas is dissolved in oil during the compression process? Measurements on full-scale pumps are difficult to perform accurately and tests results have shown widespread data scattering. The re-dissolution process is low and fluid transfer in the pump is very fast. Gas re-dissolution is therefore often considered to be negligible within the pump. Re-dissolution is often ignored when selecting the capacity and staging of a helico-axial MPP. This is a conservative assumption in order to predict the pressure rise delivered by the pump. However, in some cases, this assumption may be non-conservative, for instance to ensure that the flowrate does not reach the lower limit in the pump.

Multiphase pumps are sometimes used in heavy oil fields produced with steam. As the temperature rise is generally small during the compression, condensation of steam can occur in the pump. Contrary to re-dissolution, steam condensation is a fast phenomenon which cannot be ignored when selecting the pump. As the steam fraction is often large within the vapor phase, the reduction of gas fraction can be significant.

2.2.3.4 Reliability and Availability

No overall statistics on MPPs reliability and availability have been published until now, or even established.

The main incidents which influence the reliability and availability of MPPs are:

- seal leakage: the sealing system has a major impact on pump reliability;
- bearing failures;
- wear of rotating or static components (sand);
- long dry-running.

2.2.3.5 Sealing

As indicated above, the composition of the multiphase mixture can fluctuate rapidly. Changes of flow composition have an impact on several mechanical parts. Components must be tolerant to the passage of long gas pockets. The pump and its drive in particular can be subjected to large torque and power variations. The axial thrust on the shaft also varies with possible consequences on the thrust bearing. Fluctuations in flow composition also have considerable influence on certain critical components such as the seals.

Multiphase pumps are manufactured with single or double mechanical seals. In mechanical seals, the two sealing faces, perpendicular to the axis of rotation, slide relative to each other. An intermediate fluid, which is usually the pumped liquid, must lubricate the faces. Mechanical seals have been used for a long time in quasi-static conditions in pump and gas compressors. In MPPs, variations of flowrates and gas fraction as well as the presence of solid particles create severe service conditions for mechanical seals.

The seal faces can be made of carbon graphite or silicon carbide. Carbon graphite seals tolerate a temporary lack of lubricant but are more susceptible to damage by solid particles. Silicon carbide seals are wear-resistant and recommended if sand is present, but they require permanent lubrication that is hard to guarantee with single seals.

When double mechanical seals are used, a clean intermediate barrier fluid is used, generally the lubricant fluid. The intermediate fluid is pressurised slightly above the internal pressure to avoid leakage of the process mixture and to lubricate the seal faces with the clean fluid. The differential pressure must be limited to avoid the loss of lubricant in the process fluid.

Double mechanical seals are highly recommended when leakage through the seal can create a toxic or explosive atmosphere, such as when the pumped mixture contains H_2S or when the pump is located in an enclosed area or container.

2.3 POSITIVE DISPLACEMENT PUMPS

Two main positive displacement pump concepts are used in multiphase flow: twin-screw pumps and progressing cavity pumps. They are both included in the sub-category of rotary pumps, named after the basic motion that ensures the transfer of fluids.

In single-phase flow, these two types are particularly commonly used in the oil industry, particularly with viscous products. They offer the advantage of continuous delivery of fluids, without flow pulsing, as the fluid is transferred by a continuous rotating movement. They have also good flow capacity, up to 1,500 m^3/h and more for screw pumps and 200 m^3/h for progressing cavity pumps.

2.3.1 Twin-Screw Pumps

Twin-screw pumps have been used in industry for more than 70 years in single-phase flow applications, mainly to transfer fluids of high viscosity or at low suction pressure (or low available NPSH) when centrifugal and other pumps are unpractical. In the oil industry, they are used especially with heavy crude oil. However, they are relatively heavy and expensive.

Twin-screw MPPs can cover total volumetric flowrates (oil, water & gas) at suction conditions from 10,000 bbl/day up to 300,000 bbl/day (1,900 m^3/h), depending on their size. They can accept differential pressures of up to 70 bar (1,000 psi) but the main manufacturers propose designs up to 100 bar of differential pressure.

2.3.1.1 Principle and General Arrangement

A twin-screw pump is composed of two parallel helical screws which engage with each other. The helical channel formed by each screw is periodically obstructed by the other screw. This arrangement delimits small chambers, or locks, filled with the multiphase mixture. The chambers moves continuously along the axes when the screws rotate and they transfer the fluid from suction side to discharge side (Figure 2.5). The theoretical volumetric flowrate of the pump is governed by the volume of the chambers and thus by the size and pitch of screws and their rotation speed. One shaft is directly coupled to the drive; the second one is driven through a timing gear which transfers the torque and synchronises rotation. The gear is usually installed at the opposite side of the drive end. The screws usually rotate between 1,500 and 2,400 rpm.

The shafts and screws can be made of several pieces or of one piece. Using several pieces offers the advantage of being able to replace only the screws when they are worn. However, this is not always an advantage. With a single machined piece for each rotor, the screws are less prone to fatigue cracking and dynamic balancing is easier. Each rotor also has a greater stiffness and shaft deflections are lower. As explained below, shaft deflections are an important issue with respect to clearances.

Usually, a pair of screws is arranged back-to-back on each shaft, as shown in Figure 2.5, to balance the axial thrust produced by the pressure. In this configuration, the high-pressure

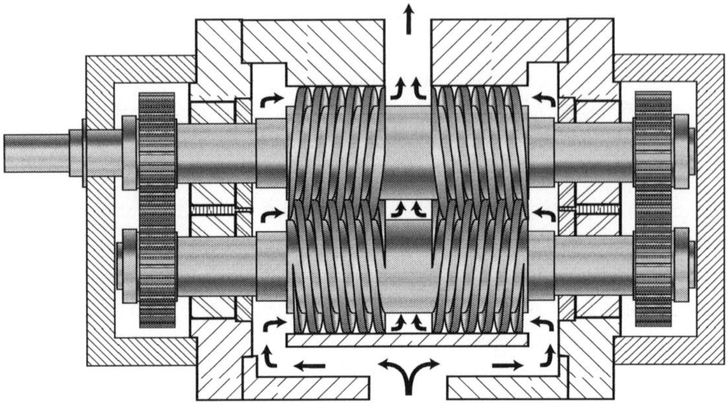

Figure 2.5

General arrangement of a twin-screw pump

discharge is placed in the middle of the shaft and the low-pressure suction is placed on the two sides. The seals are subjected to a lower differential pressure. This layout is reversed in some twin-screw pumps.

To allow relative motion in multiphase applications, the two screws are not in contact, nor are they in contact with their housing. This is a major difference with single-phase twin-screw pumps where contact is allowable when the process fluid provides the necessary lubrication. This cannot be guaranteed in multiphase flow as the pump can run dry during gas pocket flow. In the clearance gaps, the differential pressure generated by the pump creates leaks, generally called backflows, which reduce the theoretical volumetric capacity of the pump, and thus its volumetric efficiency. The overall pump capacity is principally governed by backflow between the first chamber and the suction duct.

Mechanical clearance plays an important role in screw pumps. Very tight clearances reduce fluids leaks and improve pump capacity and efficiency. However minimum clearances must be guaranteed to enable thermal expansion and shaft deflection under pressure loads. In this respect, the back-to-back arrangement of screws requires longer shafts with greater flexibility and consequently, larger gaps.

The number of chambers along the screws varies according to the manufacturer. This number directly influences the distribution of internal pressure and the volumetric losses. It must be sufficient to ensure good volumetric efficiency, but only to a certain limit as a shaft that is too long requires wider gaps with increased leakage.

Brandt and Wietstock [1992] have presented experimental results on pressure distribution and explained its influence on pump volumetric efficiency. Vetter and Wincek [1993] have given formulas for leak flows versus differential pressures in the three types of clearances: between the screw outside edge and the housing, between the outside of one screw and the shaft of the other one, and between their sides.

For 100% liquid (incompressible flow), the backflow entering a chamber is identical to the flow leaving it. A uniform backflow with same pressure losses is observed all along the screw. The total differential pressure is distributed equally between the chambers. With high-viscosity liquids, which was the conventional application of twin-screw pumps, the relationship between differential pressure and backflow is practically linear as the flow in the clearances is essentially laminar. In multiphase applications, the viscosity is often low, and clearances are also made wider to avoid any unlubricated contact in the presence of gas. The flow in the clearances is then generally turbulent with a quadratic relationship between backflow and differential pressure [Vetter *et al.*, 2000].

In a compressible flow such as a multiphase mixture, a fraction of the backflow entering the chambers is used to compress the fluid. The pressure and backflow distributions are no longer evenly distributed along the screw. Backflows are smaller at the screw inlet than the outlet. Consequently, for a given total pressure drop, the reduction of pump capacity, governed by the backflow on the suction side, is lower with a compressible fluid than with an incompressible one. The pump volumetric efficiency increases with the gas fraction (see Figure 2.6). This advantage has also a counterpart: the pressure forces acting on screws are translated towards the middle of the shaft. Deflections are thus greater in two-phase flow than in single-phase flow for the same differential pressure and the gaps required are therefore wider. For this reason, the reverse layout with low pressure in the centre, is sometimes proposed.

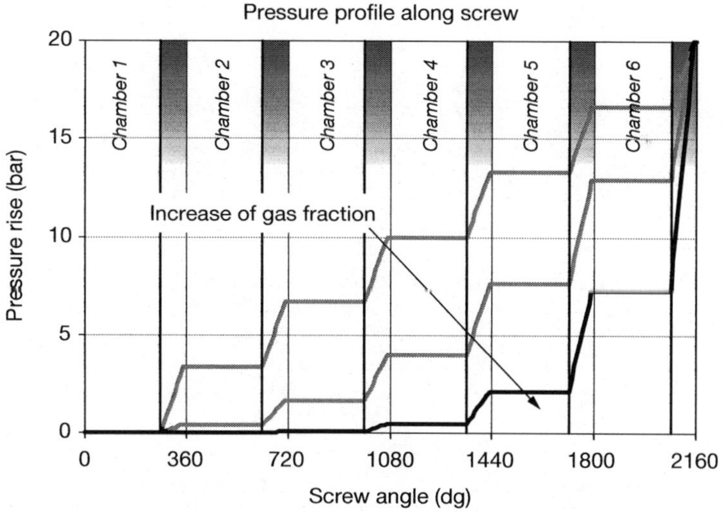

Figure 2.6

Profile of pressure in a twin-screw pump

The above behaviour is observed when there is enough liquid to fill the gaps and maintain a significant difference in pressure between the chambers. At very high gas fractions, some gas can flow through the gaps, reducing the differential pressure between successive chambers. The backflow and pressure distributions resemble the distribution obtained with an incompressible liquid, leading to a reduction in volumetric efficiency [Vetter *et al.*, 2000].

Normally, the casing of a screw pump forms an inlet capacity which can store some liquid and smooth out short transient gas slugs. However, in the case of long gas slugs with average GVF higher than 95-96%, an external liquid recycle system must be added to maintain the internal seal and pump cooling. For instance, the liquid storage and recycle system shown on the right of Figure 2.7 is sufficient to ensure pumping with a gas pocket for approximately 30 minutes.

The inlet casing space may also become a disadvantage if sand is produced in the well stream. It can become a sand trap and feed the screws with a high sand content fluid, with no means of flushing the pump to clean it, other than dismantling.

Resistance to wear in the presence of solid particles (sand) is obtained by choosing a special screw profile and an adequate hard coating.

Figure 2.7

Twin-screw pump with liquid recycle system (on the right); Leistritz Pumpen GmbH

2.3.1.2 Typical Duties, Performance

Twin-screw pumps – in the same way as all volumetric pumps – do not deliver a pressure rise, as often stated, but a constant volume of fluid at each shaft revolution, at least as long as the internal backflows can be ignored. The resulting pressure rise, however, influences the flowrate because of internal backflows. The overall response of a twin-screw pump typically has the form shown on Figure 2.8. The theoretical flowrate is only observed for zero pressure rises.

When the rotational speed of the pump is increased, the response curve is translated to the right, towards higher flowrates, and it becomes steeper because relative backflows are reduced, giving higher volumetric efficiencies.

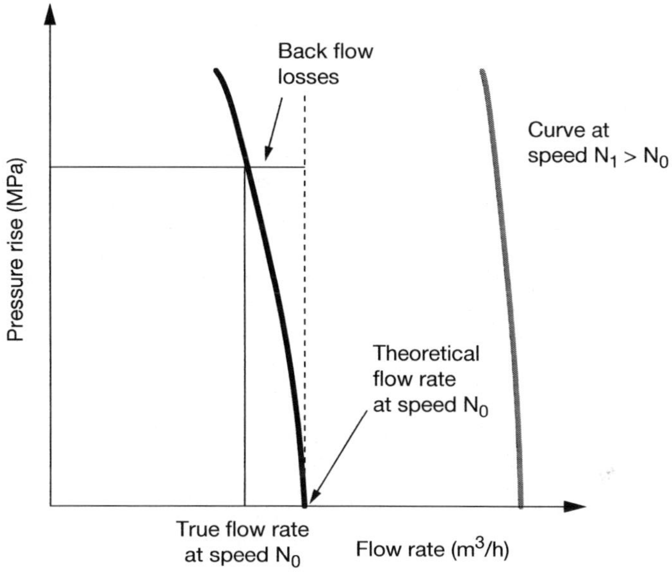

Figure 2.8

Typical performance diagram of twin-screw pump

The pressure rise depends on the resistance of the hydraulic system to the imposed flowrate. As the flow path is not continuous through the pump, the upstream and downstream systems are practically isolated by the pump. The suction pressure results from the response of the upstream conditions to the flowrate taken in by the pump (piping, well and reservoir). The discharge pressure is dictated by the response of the downstream system (piping, separator, etc.)

If a valve is closed and stops the flow on the downstream side, the discharge pressure can build-up quickly and can produce serious damage. For this reason, a pressure-relief safety valve must always be installed at the outlet of a volumetric pump. In some designs of twin-screw pumps, the relief valve is included in the pump body. If the pressure increases above the setpoint, the valve opens and the fluid is re-circulated to the inlet.

2.3.1.3 Advantages, Limitations

The main advantages of twin-screw pumps are:

- low sensitivity to variations in flow conditions (suction pressure, GVF);
- their suitability for very low suction pressure and high viscosity liquids;
- relatively low rotation speed, allowing the use of cheap and conventional bearings and drivers.

And their limitations:

- dry-running limitations requiring liquid recycle if long gas pockets are possible;
- shaft deflections under hydraulic loads;
- higher number of seals, with limited shaft clearance.

2.3.2 Progressing Cavity Pumps

Progressing cavity pumps (PCP) are composed of a screwed metallic rotor mounted inside a stator made of flexible material, generally a lined elastomer. They were invented in 1930 by a French engineer, René Moineau. The rotor can be designed with a single screw or a combination of several screws (multi-lobe pumps). The internal cavity of the stator has the shape of a multiple helix (number of rotor screws plus one). When the rotor rotates, it axis describes a circle in the opposite direction. The radius of this circle depends on the number of screws, as well as the speed of rotation of the rotor axis and the helix pitch relationship between stator and rotor. In any position, the rotor is in contact with the stator at several points to form closed cavities. The cavities progress from one end to the other one when the rotor rotates. The pump capacity is governed by the size of the chambers and the rotation speed.

The rotation speed of a PCP is limited because the rotor is subjected to a complex motion with unbalanced mass. Their capacity is consequently relatively limited (around 200 m^3/h in single-phase flow).

The discharge pressure is not determined by the characteristics of the pump but by the resistance of the downstream piping system to the imposed flowrate. The volumetric efficiency of a PCP is directly related to the size of the gaps and the number of screw revolutions (length of rotor).

The use of an elastomer material for the stator enables tight compressive contact with the rotor and avoids relying on viscosity to limit backflows. This feature is of particular interest if the viscosity can vary widely at start-up or during operation, for instance due to temperature variations or to thixotropic behaviour of the fluid. It also makes it possible to handle slurries since abrasion in gaps is reduced. The elastomer material must be selected according to the fluid temperature and composition and the possible presence of gas or solid particles.

The maximum operating temperature which can be withstood by elastomers can also be a limiting factor.

Comprehensive details on principle, design, selection and operation of PCP were given by Cholet [1997]. PCPs have long been used in multiphase conditions in wells for pumping

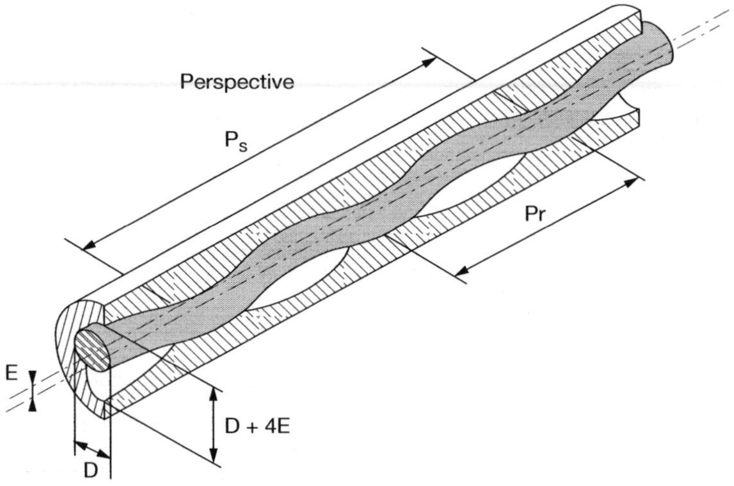

Figure 2.9

Principle of a PCP-after Cholet [1997]

mixtures at low gas fraction (lower than 50%). An overview of developments for boosting multiphase flows at high gas fractions were presented by Mirza & Wild [1997].

Basically, the working principle of progressing cavity pumps and twin-screw pumps are the same. The transfer of fluids is ensured by the continuous motion of the cavities along the shaft with progressive compression of gas by backflows between the stator and rotor. Consequently, some modifications need to be made to the standard PCP design to ensure appropriate operation in multiphase flow. In particular, the elastomer material must be improved to withstand the temperature increase produced by the compression at gas fractions exceeding 80%. Backflows must be also mastered to ensure a correct pressure and temperature profile along the shaft.

Bratu and Seince [2005] proposed a new multiphase PCP concept with calibrated orifices in the rotor. These orifices allow fluid backflows between the successive cavities. The purpose of this design is to control backflows more accurately than through clearances between stator and rotor, and thus to obtain progressive compression of gas along the rotor.

2.4 HELICO-AXIAL ROTODYNAMIC PUMPS

Unlike positive displacement pumps, centrifugal pumps do not impose displacement on a finite volume of fluid, but they perform a continuous transfer of energy from the prime mover to the fluid. During this process, mechanical work is firstly transferred to the fluid in a rotating part of the machine and then converted into pressure in a static part.

Centrifugal pumps normally entail a head loss when gas is present in the liquid, with a rapid drop in performance above a few % of gas (typically 5-8%). A special pump concept, called "helico-axial" avoids this behaviour.

2.4.1 Principle and General Arrangement

Multiphase helico-axial pumps are multistaged rotodynamic pumps (Figure 2.10). Each stage is composed of a rotating part mounted on a shaft, the impeller, and a static diffuser. The number of stages depends on the required head. So far, the highest number of stages for surface pumps is 15 and this is only limited by the dynamic behaviour of the rotating assembly. Down-hole pumps, composed of several assembled modules may have a greater total number of stages.

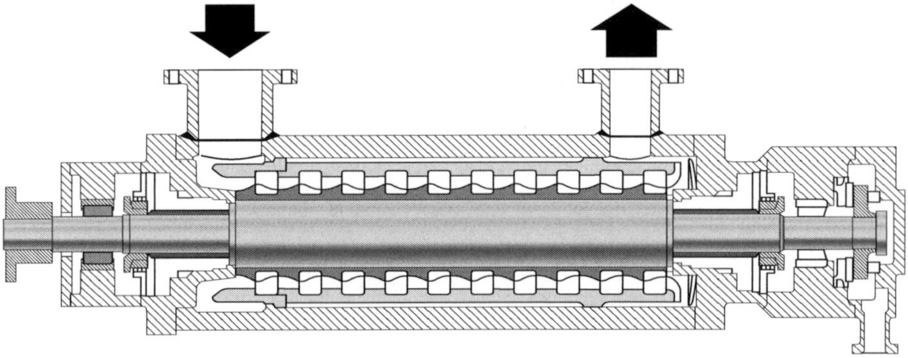

Figure 2.10

Helico-axial multiphase pump (Sulzer Pumps mechanical concept)

A typical helico-axial compression stage is shown in Figure 2.11. The name "helico-axial" derives from the combination of the helical shape of the blades and the global axial flow direction, *i.e.* parallel to the shaft. The free open hydraulic channels accommodate solid particles in the flow but care must be taken to prevent them accumulating in the pump casing or gaps.

Until now, helico-axial multiphase pumps have been manufactured with impeller diameters ranging from 75 mm to more than 400 mm. Rotation speeds are usually chosen between 3,500 rpm and 6,500 rpm.

The Poseidon helico-axial concept covers total volumetric flowrates (oil, water and gas) at suction conditions ranging from 22,000 bbl/day (146 m^3/h) up to 550,000 bbl/day (3,700 m^3/h). The largest pumps can rate up to 6 MW: helico-axial pumps are the largest multiphase pumps manufactured in the world.

To date, helico-axial pumps in service cover a wide range of operational parameters including low suction pressures, high GVF up to 93% on average, *i.e.* gas to liquid ratio (GLR) up to 15. Actual gas fractions flowing through the pump can vary from 0 to 100% and several pumps have already been run on 100% gas with no problems for several days as a result of process trips at weekends on unattended fields.

The gas volumetric flowrate is reduced during the compression of the multiphase mixture, in the same way as in compressors, as illustrated in Figure 2.12. The GVF is reduced and the average fluid density is increased. For this reason, the pump is generally equipped

Figure 2.11

Helico-axial multiphase compression cell

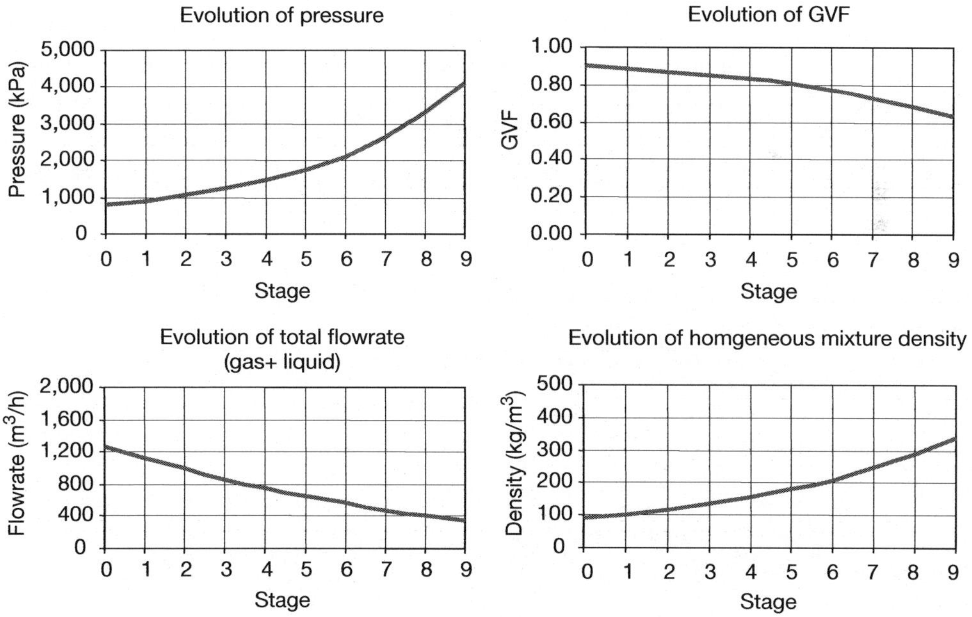

Figure 2.12

Typical profile of pressure, total flowrate, GVF & mixture density in a helico-axial pump with several sets of different stages, the size and shape of which provide adjustment for the decreasing volumetric flowrate. Owing to the large operating range of each stage, the size does not have to be changed at every stage, as with centrifugal compressors, but only every three, four, five or more stages, depending on the mixture compressibility.

2.4.2 Duties, Performance

2.4.2.1 Head and Efficiency

The specific work ΔW (energy per unit of mass flow) transferred to the fluid by a centrifugal pump is usually characterized by the head $H = \Delta W/g$ which has dimension of length.

The use of the head derives from historical hydraulic concepts. The head depends on the pump geometry, on its rotation speed and also on the fluid flowrate, but is independent of fluid density. The difference between the discharge and suction pressures is not a characteristic of a pump: it depends on the fluid density and can be determined from the head by using the Bernoulli equation (conservation of mechanical energy):

$$H = \Delta[z + p/\rho g + V^2/2g] \tag{2.1}$$

The head of a multiphase mixture is difficult to characterize because it depends on the velocity slip between the phases. As the slip is difficult to measure, the true head of a flowing mixture is difficult to assess, especially within the pump. Usually, a no-slip condition is assumed for estimating pump performance.

In the same way as any centrifugal pump, a helico-axial MPP basically delivers a head. The head is a function of the fluid velocity flowing in the pump, *i.e.* of the volumetric flowrates. The differential pressure between suction and discharge also depends on the multiphase mixture density.

The energy transferred by each impeller to the fluid, per unit mass, is given by the relationship:

$$\varepsilon_c = U_2 V_{U2} - U_1 V_{U1} \tag{2.2}$$

where U is the impeller circumferential velocity, and V_U the component parallel to U of the absolute fluid velocity in a fixed reference frame. Suffix 2 and 1 refer respectively to the outlet and inlet of the impeller. The corresponding head is

$$H_c = \frac{1}{g}\left(U_2 V_{U2} - U_1 V_{U1}\right) \tag{2.3}$$

The overall energy transferred by the pump is cumulated over the number of stages. The above relationships are derived from the Euler theorem which holds for a perfect fluid (no viscosity effects) and perfect transfer of energy. As this is not the case, the actual energy transferred to the fluid in single-phase flow is lower and usually derived from the ideal conditions with an efficiency multiplier η_i:

$$\varepsilon_{SP} = \eta_i\left(U_2 V_{U2} - U_1 V_{U1}\right) \qquad H_{SP} = \frac{\eta_i}{g}\left(U_2 V_{U2} - U_1 V_{U1}\right) \tag{2.4}$$

The multiplier η_i describes the efficiency of mechanical energy transfer from the shaft to the fluid. Some other energy losses must be added, mainly occurring in bearings and seals, in order to obtain the overall external efficiency of the pump as seen by the driver.

2.4.2.2 Multiphase Performance Multipliers

The operational ability of a multiphase pump is usually quantified by comparing the delivered head in two-phase flow H_{TP} to the head in single-phase flow H_{SP} at the same flowrate, equal to the total of liquid and gas flowrates. This ratio E_H is called the "multiphase head multiplier" or sometimes "efficacy":

$$E_H = \frac{H_{TP}}{H_{SP}} \tag{2.5}$$

Similarly, an efficiency multiplier can be defined as the ratio of efficiency in two-phase flow to the efficiency in single-phase flow (or compression power ratio):

$$E_\eta = \frac{\eta_{TP}}{\eta_{SP}} \tag{2.6}$$

With a compressible mixture, the GVF, pressure and flowrate often vary significantly as the flow moves through the pump. These parameters influence pump performance and the multiphase multipliers. To be accurate, the multiphase multipliers must be calculated for every stage. A global multiplier simply provides an average value that depends on the number of stages, and it is difficult to correlate with the flow conditions.

The head H_{TP} is a conventional or apparent head assuming that the multiphase mixture behaves like a homogeneous single-phase fluid, without velocity slip. Generally, other assumptions are also made, like thermal equilibrium between the phases or isothermal flow. H_{TP} is related to compression energy transferred to the mixture and to the differential pressure measurements by the relationship:

$$H_{TP} = \frac{1}{g}\left[\frac{(1-\alpha)\rho_l}{(1-\alpha)\rho_l + \alpha\rho_g}\frac{\Delta p_{TP}}{\rho_l} + \frac{\alpha\rho_l}{(1-\alpha)\rho_l + \alpha\rho_g}\int_{p_i}^{p_i+\Delta P_{TP}}\frac{dp}{\rho_g(p,T)}\right] \tag{2.7}$$

where α is the GVF, p_i the stage inlet pressure, ρ_l and ρ_g the liquid and gas densities. In this equation, the compression energy per unit mass of mixture is the sum of the compression energy of the liquid and gas phases.

The integral can be calculated if the evolution of temperature T is known during the compression process. This calculation requires performance of a heat balance, *i.e.* to solve an energy equation. This is detailed below, in the thermodynamic section.

To avoid this calculation, two simplifications are performed. The phases are generally assumed to be in thermal equilibrium and the fluid behaviour is often assumed to be isothermal. As shown below, the second assumption is generally acceptable, except if the GVF and the compression ratio are too high. In the latter cases, the behaviour can be approximated by a polytropic transformation.

In general, the true mixture density, the phase velocities and the thermal behaviour are not known. They have an impact on the true head, but this has no consequence on prediction of performance because their effects are implicitly included in the multiphase multipliers E_H and E_η which are determined experimentally.

Tests on a multiphase loop allow the influence of a series of parameters on the multiphase efficiency to be determined: local fluid conditions (phase densities and flowrates, GVF, pressure), hydraulic geometry and pump rotation speed.

Every data point measured in known flow conditions is processed according to the above equation to calculate E_H. The corresponding head in single-phase flow is calculated from the pump response in similar flow conditions. According to the above definition, multiphase efficiency is unity in a single-phase liquid (GVF = 0) as well as in dry gas (GVF =1.0) if the temperature rise produced by the gas compression is correctly taken into account. Between the two extreme cases, multiphase efficiency is lower than 1, enlightening a deviation from the theoretical homogeneous fluid in thermal equilibrium, resulting mainly from phase slip. Figure 2.13 shows the evolution of typical multiphase efficiencies versus GVF and density ratio of gas and liquid. The curves depend on the geometry of the hydraulic stage and thus depend on the pump.

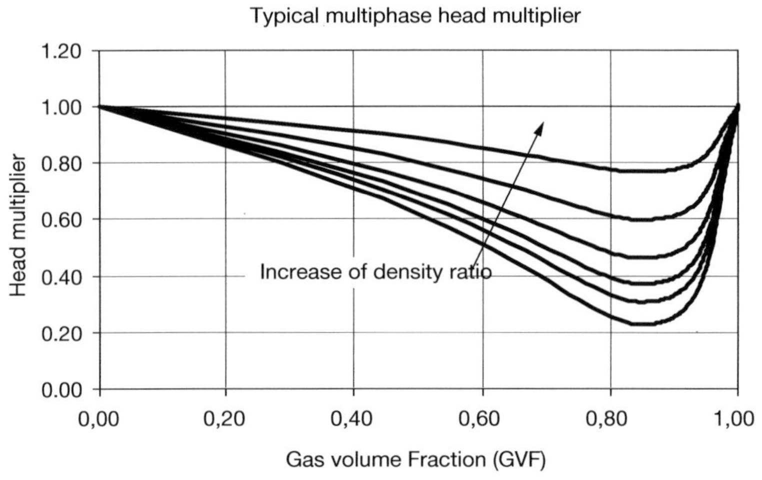

Figure 2.13

Typical head multipliers in two-phase flow (as a function of parameters: GVF and density ratio)

All the data processing described above is performed by the pump manufacturer or designer and is described here for general information. The pump user does not become involved in the detailed fluid mechanics of multiphase pumps and is provided with performance curves (or characteristic curves) which describe the pump response in steady-state operation. The pressure rise (or head) and efficiency (or power) are plotted versus the flowrate at constant speed. Curves are often given for several rotation speeds.

In multiphase flow, the GVF and suction pressure (at low suction pressure) influence the characteristic curves. For this reason, a set of curves is often plotted for several GVF or several suction pressures, as shown in Figure 2.13. Other information such as best efficiency locus, minimum flow limit, etc, is also sometimes given on the diagrams.

2.4.2.3 Characteristic Curves

As indicated above, the head depends on the flowrate. The characteristic curves of a pump describe the variation of head and efficiency versus flowrate (Figure 2.14). For a given rotation speed, optimum efficiency in energy conversion is obtained for a certain flowrate, called the best efficiency point (BEP).

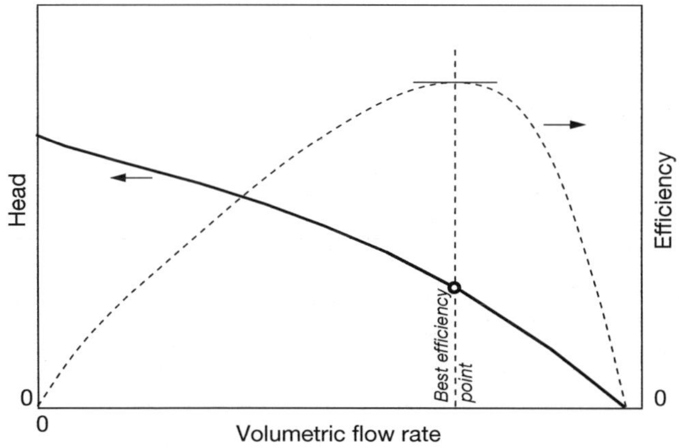

Figure 2.14

Characteristic curves (head and efficiency)

In multiphase flow, the GVF influences the characteristic curves and, for helico-axial pumps the suction pressure at low pressure. For this reason, a set of curves is often plotted for several GVF, as shown on Figure 2.15. Other information, such as best efficiency locus, minimum flow limit, etc, is also sometimes given on the diagrams.

Flowrates used on the diagram abscissa are either liquid flowrates or total flowrate. Gas flowrates could also be used, but this is less frequent. In steady-state conditions, the same information is contained in the curves, whatever the flowrate. However care must be taken to use the right flowrate for interpreting the curves. If total volumetric flowrates are used with customary oil units, liquid flowrates and gas flowrates must be converted in a common unit before adding them, since they are usually expressed in two different units, in barrels per day and cubic feet per day respectively. At high speed and high compression ratio, characteristic curves of helico-axial pumps can be as steep as curves of volumetric pumps.

Operation at low flowrate must be avoided for several reasons: possible vibrations due to internal flow re-circulations, high temperature rises and possible surge behaviour as in a compressor (flow pulsation regime).

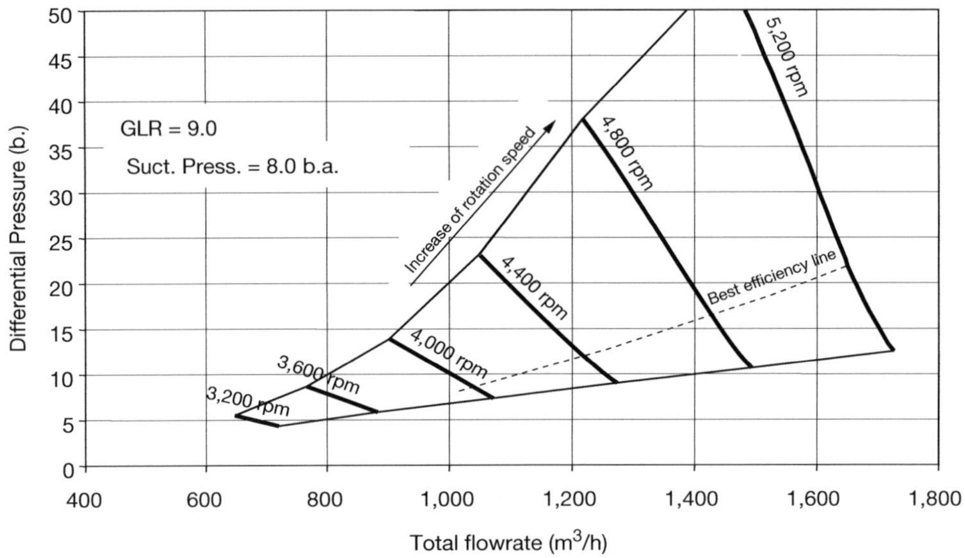

Figure 2.15

Example of characteristic curves at 90% GVF for a large capacity pump (GLR = gas liquid ratio in suction conditions)

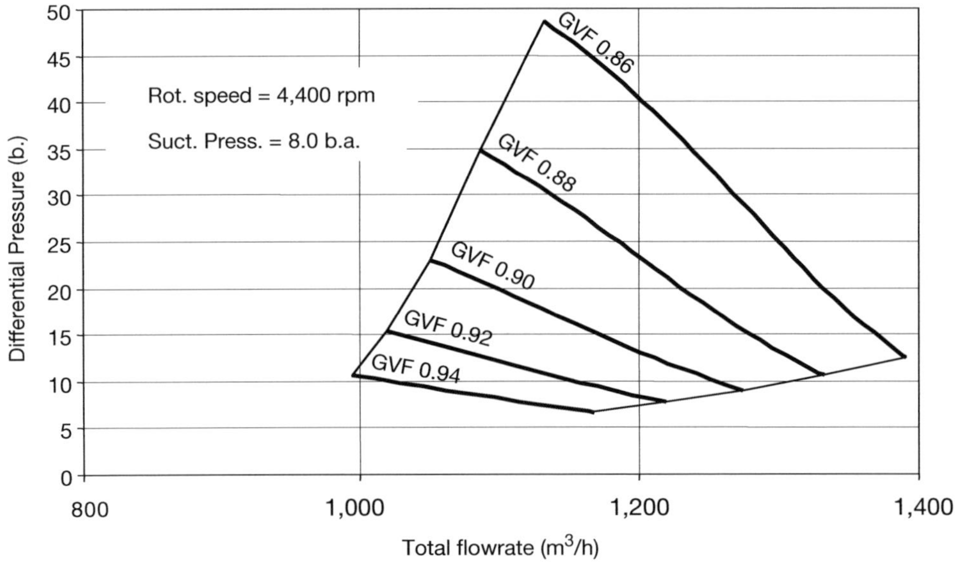

Figure 2.16

Example of characteristic curves at constant speed and variable GVF (same pump as in Figure 2.15)

2.4.2.4 Affinity Laws

Similar impellers, *i.e.* having the same geometric shape and proportions but different sizes, are characterised by a dimensionless number, the specific speed N_s:

$$N_s = \frac{NQ_{BEP}^{0.5}}{H_{BEP}^{0.75}}$$

where N is the rotation speed, Q_{BEP} the flowrate and H_{BEP} the head at BEP. The head must be expressed in units of specific energy and the other two quantities in coherent units to obtain a true dimensionless number independent of the unit system (such as N in rad/s, Q_{BEP} in m^3/s and H_{BEP} in J/kg). In practice, the head is often expressed in length unit and the rotation speed in revolutions per minute and thus, when a specific speed is given, it may depend on the unit system used. Nevertheless, the specific speed characterises the performance of a series of similar impellers.

Affinity laws are applicable in multiphase flow, provided that the flow conditions are also similar (same GVF for instance, same density ratio between phases).

In single-phase flow, the best efficiencies are obtained when the general shape of the impeller is suited to the specific speed required by the pump duty. For instance, axial flow impellers are best-suited to high specific speed and radial flow impellers to low specific speed. In multiphase flow, the shape of the impeller is dictated by the search for a good performance multiplier. The helico-axial shape is best-suited to high volumetric flowrates. This has an impact on the basic efficiency of small pumps.

2.4.2.5 Multiphase Performance Models

The velocity slip which exists between the gas and liquid phases explains the observed head drop. Under the acceleration forces the liquid is accelerated and flows faster than the gas. As a consequence of the velocity slip, the true gas content is higher in the pump than in the incoming mixture. The gas expansion in the low-pressure zone (especially near the impeller suction side) amplifies the increase in gas content and liquid acceleration. As long as the flow remains well dispersed (bubbly flow) the head drop is limited by the interactions of gas and liquid (drag and inertia forces). But the transverse component of the acceleration field quickly involves the separation of phases and this reduces the momentum transfer between the phases. The head drop mechanism has been modelled by several authors: de Crécy [1983, 1986], Homer [1985], Furuya [1985].

The good performance of the Poseidon pumps in two-phase flow is linked to the special shape of the impeller. This shape limits centrifugal accelerations and avoids low-pressure zones. The overall shape is designed to limit transverse acceleration and thus phase separation. The shape also reduces velocity slip and thus the increase in true gas fraction. Calculations performed with one of the models mentioned coincide closely with the experimental measurements [Coste and Vilagines, 1993].

2.4.2.6 Flow Instabilities

Centrifugal pumps can be subject to flow instabilities in multiphase flow at low flowrates. In some two-phase flow conditions, the pump and piping system can build up cyclic fluctuations of pressure and flow, at frequencies from 1 Hz to a small fraction of a Hz. These instabilities are also called surge or heading and appear similar to the surge in centrifugal compressors. Even though these instabilities are less destructive in MPPs than in compressors, it is inadvisable to leave an MPP running in surge conditions because flow fluctuations are associated with strong mechanical vibrations that can break bearings or induce fatigue damage.

The origin of the pressure instabilities in two-phase flow has yet to be fully clarified and no general theory has been proposed. Since the fluid mixture is compressible, the phenomenon can be associated with a local positive slope of the pump's characteristic curve, as with surging of a compressor, but some authors also associated it with a change in flow regime in the pump, or with internal flow recirculation.

Pessoa and Prado [2001] presented a summary of observations and empirical criteria for the onset of instabilities in conventional centrifugal pumps used in oil wells (electro-submersible pumps – ESP). However these data cannot be generalised. As indicated by the authors and previously by Esson and Cohen [1996], pressure instabilities are heavily dependent on the impeller shape and thus on the pump.

Helico-axial pumps are less prone to these phenomena and they can be managed if the risk of occurrence is well evaluated. The stable flow domain of helico-axial impellers is much broader than given for conventional pumps. For this type of impeller, Esson and Cohen presented a more realistic boundary for stable flow, expressed as a relationship between GVF and flowrate relative to the best efficiency (BEP) flowrate. Experiments have shown that, for a GVF greater than 50%, the stable domain of Poseidon pumps is broader than that observed by Esson and Cohen. The flowrate can be reduced to 55% of the BEP flowrate of each impeller.

Concerning gas locking, Pessoa and Prado [2001] observed that the phenomenon appears after pressure instabilities but it is difficult to characterise because the concept is not clear. Indeed, the usual definition of gas locking appears to depend on the experimental system. The gas fraction does indeed have an influence on the pump performance and, depending on the piping system, can lead to a flow blockage or not. This is further discussed in the "multiphase pump operability" section.

2.4.3 Advantages, Limitations

The main advantages of helico-axial hydraulic pumps are:

- their ability to pump any GVF from 0 (100% liquid) to 1.0 (100% gas) on a continuous basis;
- mechanical simplicity and reliability (single shaft, few seals, rotodynamic principle);
- self-adaptation to flow changes;
- high possible pressure rise (up to 150 bar and above with recent high boost designs);

- ability to accept very high suction pressures (the design pressure of the whole pump – suction and discharge flanges – is generally the maximum allowable working pressure of the downstream pipeline);
- compactness;
- tolerance to solid particles (open type channel, without tight clearances);

while their drawbacks are:

- power penalty at high fluid viscosity (but predictable performance achieved for oil viscosities as high as 250 cP, and successful operation also observed at 2,500-3,000 cP in cold conditions);
- unsuitability for low flowrates (possible flow instabilities).

2.5 MULTIPHASE PUMP OPERATION

2.5.1 Pump Duty

2.5.1.1 Definition of Operating Domain

The main parameters which influence the selection of an MPP are:

- the gas and liquid flowrates, or a flowrate, a gas fraction (GVF) and a water fraction (or water cut);
- the fluid densities (gas and liquid);
- the required pressure rise;
- the suction pressure and fluid temperature;
- the oil viscosity at least.

These parameters must be given, or calculated, for suction conditions. The gas-liquid ratio (GLR) in suction conditions is also used instead of the GFV.

In order to optimise selection of the type of pump, the materials, the sealing and bearing components, additional information on operational factors are also required:

- fluid characteristics and gas composition (presence and concentrations of corrosive components, possible wax deposition, specific heat coefficients if temperature rise can be a problem);
- possible presence of solid particles (sand, scale formation).

The gas content is often characterised by the gas volumetric fraction GVF which is the ratio of the actual gas volumetric flowrate to the total flowrate in suction conditions. The gas-liquid ratio GLR in suction conditions is also used instead of the GVF. The following relationships hold between GVF and GLR:

$$GLR = \frac{GVF}{1 - GVF} \qquad GVF = \frac{GLR}{1 + GLR} \tag{2.8}$$

The GLR must not be confused with the gas oil ratio GOR, which describes the relative volume of gas and oil produced once the two fluids are brought to standard conditions.

As already mentioned, it is often difficult to specify a single duty point in oil production due to flow variations on several time scales. Short-term variations in the operating point and hydrodynamic phenomena such as slug flow, are usually handled by the natural flexibility of the pump or, if required, by a flow homogeniser, a recycle loop or a control system with speed variation to increase the possible operating domain. Medium-term variations are handled by the flow homogeniser, recycle loop or control system with speed variations of the pump.

The above parameters are often given with a range of possible variations or a profile of evolution versus time. If the variation is large, speed variation may be insufficient to cover the anticipated operating domain. After some time, the pump may be retrofitted to adjust its characteristics to actual production, by modification of the screws (for twin-screw pumps) or restaging (for helico-axial pumps).

2.5.1.2 Pump Selection Procedure

The selection of an MPP for an application usually begins with a preliminary approach focusing on the main duty point.

The first step consists in selecting the type of pump (progressing cavity, twin-screw, helico-axial or others). This is usually done mainly on the basis of the total volumetric flowrate (gas plus liquid) with certain additional criteria taken into account: suction pressure and pressure rise, liquid viscosity, type of system curve. Two types are often pre-selected for assessment on a cost-competitive basis.

The second step consists in selecting a pump model from the manufacturer documentation. Manufacturers generally propose a range of standardised pump sizes based primarily on the total flowrate. At this step, it is possible to verify that the required pressure rise is within the capability of the selected pump model and to estimate the driver power. This requires an estimate of the efficiency from the manufacturer.

The third step consists in finalising the pump design to optimise the standard model for the considered application, for example to adjust the number of stages, select the nominal rotation speed, perform a predictive calculation of performance, etc. In general, this step must be performed by the manufacturer. Definition of the pump operating domain is one of the outputs of this step. It is then possible to verify that this domain matches the anticipated evolution of production with time, and to determine whether or not speed variation is required.

2.5.2 Steady-State Performance Analysis

2.5.2.1 Characteristic and System Curves

In steady-state conditions, pump performance is usually described by a set of curves which provide the pressure rise (or head) and efficiency (or power) versus the flowrate at constant speed. Curves are sometimes given for several rotation speeds. When a single speed is available, affinity laws enable the adequate data to be derived. A typical example of characteristic curves is given in Figure 2.14.

The pump must deliver a pressure rise which boosts the flow in the production system. The required pressure rise depends on the flowrate and can be calculated from the difference in responses of the downstream system (export pipeline plus receiving separator) and the upstream system (piping, well and formation). The resulting curve is the system characteristic. The system characteristics generally have a positive slope since pressure losses in piping, valves, chokes, etc., increase with flowrate. In multiphase flow, this is not always true due to the influence of the liquid hold up.

The system curve is given by the difference between the discharge pressure required to boost the fluid at a certain flowrate in the downstream pipeline and the available pressure at suction for that flowrate in the pump. The pump operating point is located at the intersection between the pump characteristic curve and the system characteristic curve. If the system characteristic curve is modified (addition of a well, restriction of flow by wax or scale deposition, increase of water cut, etc.), the operating point changes. If the characteristic curve of the pump is modified (change in rotation speed), the operating point also changes. This is true for any type of pump, including positive displacement and helico-axial.

Pumps are usually designed for a given duty point on the system curve.

Flow stability requires that the slope of the pump curve is negative at the intersection point. This is always the case with a positive displacement pump. The slope can be positive with a centrifugal pump if the curve intersection is located at a low flowrate, 50 to 60% below the best efficiency flowrate.

Figure 2.17 shows an example of characteristic curves and system curves for a pump designed for deepwater applications. The system curves are plotted for two different water cuts. Without boosting (no differential pressure), the natural flowrate in the left curve situation (water cut 80%) would be 210 m^3/h. When the pump is running at 3,600 rpm, the flowrate will be increased to 410 m^3/h, and practically double.

The system curve intersects the zero flowrate axis at a positive differential pressure if the pump had to overcome a positive static head. This situation may occur when the outlet of the pipe system is located at a higher level than the inlet.

If the system curve intersects the zero flowrate axis at a positive differential pressure above the differential pressure generated by the rotodynamic pump in 100% gas, the flow can be stopped. This instability is often called gas-locking.

In oil production, however, the natural pressure of the reservoir is often sufficient to overcome the static head. In such cases, the well flows without pumping, at a certain rate. This situation does not lead to potential problems with rotodynamic pumps in multiphase flow, even with passage of long gas pockets leading to a reduction in the differential pressure delivered by the pump. The well mixture still flows, at a lower rate, and the initial performance is restored when the gas pocket is exhausted.

The system curve must be considered in order to ensure good pump performance. A control system must be installed to avoid gas-locking if the natural pressure of the reservoir is not sufficient to overcome the minimum static head at zero flowrate. Such a system can consist of liquid re-circulation (see section 2.5.5). If possible, the pump should preferably be located lower down, for example in the well (multiphase ESP).

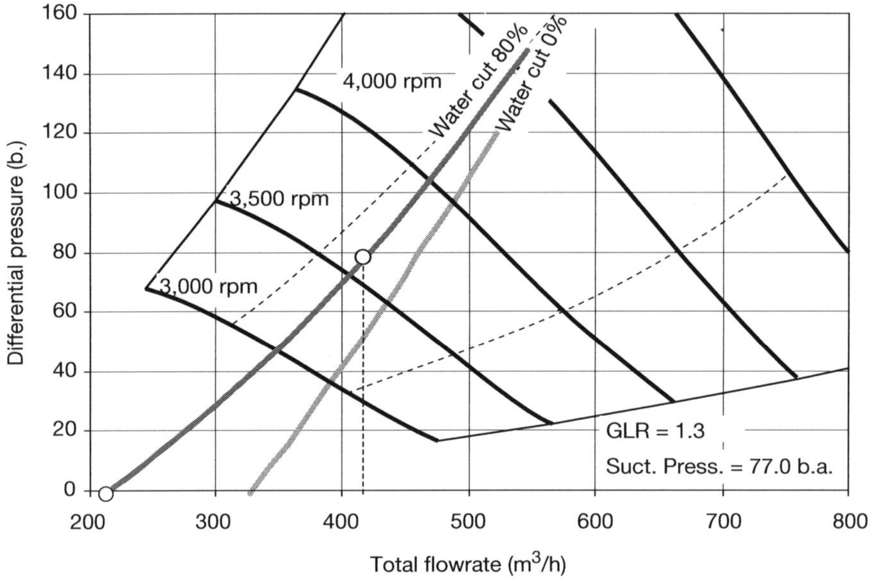

Figure 2.17

Characteristic and system curves – operating point

2.5.2.2 Parallel and Series Operations

Parallel operation of volumetric pumps does not raise specific problems. However connection of volumetric pumps in series must be approached with care. Theoretically, the use of two pumps enables a degree of energy saving, since the capacity of the downstream pump can be reduced.

As the flow is compressible, the volumetric flowrate is reduced at the discharge of the upstream pump. The inlet volumetric flowrate of the downstream pump must be set accordingly. If this flowrate is too low, the intermediate pressure will increase dangerously. If this flowrate is too high, the intermediate pressure will decrease. In practice, the intermediate pressure is unstable and the rotation speed of one pump must be controlled.

2.5.3 Thermodynamic Topics

The following developments are derived from the general principles of thermodynamics and they apply to all types of MPP.

2.5.3.1 Compression Work

The compression work W_c per unit mass of the fluid mixture is the sum of the compression work of the liquid and gas phases weighted by their respective mass fractions:

$$W_c = \frac{(1-\alpha)\rho_l}{(1-\alpha)\rho_l + \alpha\rho_g}\frac{\Delta p}{\rho_l} + \frac{\alpha\rho_g}{(1-\alpha)\rho_l + \alpha\rho_g}\int_{p_i}^{p_i+\Delta p}\frac{dp}{\rho_g(p,T)} \tag{2.9}$$

where α is the GVF, ρ_l and ρ_g the liquid and gas densities (in suction conditions for the gas density), p_i the suction pressure and Δp the pressure rise in the pump. In this expression, the liquid is assumed to be incompressible. The integral may be calculated if the gas evolution is known during the compression process. This requires solving an energy equation to determine the temperature rise.

The compression of an ideal gas is often modelled by a polytropic transformation: $p = C\rho^k$. Such a transformation generally implies an exchange of heat with the gas and the rest of the universe which, in multiphase flow, includes the liquid phase. Calculation of the compression work is straightforward and gives the following results:

$$W_{c\ poly} = \frac{(1-\alpha)\rho_l}{(1-\alpha)\rho_l + \alpha\rho_g}\frac{\Delta p}{\rho_l} + \frac{\alpha\rho_g}{(1-\alpha)\rho_l + \alpha\rho_g}\frac{p_i}{\rho_g}\frac{k}{k-1}\left[\tau^{\frac{k-1}{k}} - 1\right] \tag{2.10}$$

where $\tau = 1 + \dfrac{\Delta p}{p_i}$ is the compression ratio of the pump. The corresponding temperature rise is given by the equation:

$$\Delta T = T_i\left[\tau^{\frac{k-1}{k}} - 1\right] \tag{2.11}$$

where T_i is the inlet temperature. A thermal balance, detailed below, must be performed to determine ΔT or the polytropic coefficient k.

In an MPP, the liquid usually cools the gas down sufficiently and the temperature rise is relatively low or moderate. Significant temperature rises are only observed when a very high GVF mixture is pumped. The polytropic coefficient k is generally lower than the isentropic coefficient γ of the gas, equal to the ratio of heat capacities at constant pressure and volume: $\gamma = C_p / C_v$.

At low GVF and low compression ratio, gas evolution is almost isothermal and this assumption can be used to evaluate the compression work. The limiting case, when $k \rightarrow 1$, corresponds to the isothermal case:

$$W_{c\ isot} = \frac{(1-\alpha)\rho_l}{(1-\alpha)\rho_l + \alpha\rho_g}\frac{\Delta p}{\rho_l} + \frac{\alpha\rho_g}{(1-\alpha)\rho_l + \alpha\rho_g}\frac{p_i}{\rho_g}\ln\tau \tag{2.12}$$

Closed-form analytical formulas do not exist for a non-ideal gas. Approximate formulas using an average compression factor z between inlet and outlet can be found in textbooks on gas compressors, such as the GPSA Engineering Data Book [1998].

2.5.3.2 Temperature Rises

Calculation of the compression work involves determining the temperature rise in the pump or the coefficient k of the polytropic transformation. This calculation is also useful for design purposes as the temperature rise governs the thermal expansion and clearances between moving and unmoving parts in the pumps. The maximum temperature also influences the material selection, the choice of lubricant, the seals and piping, etc.

Temperature measurements in MPP tests show that the best temperature rise estimates are obtained with following assumptions:

- the gas is cooled down by the liquid (heat transfer),
- in addition, the mixture is heated by the internal power losses from the pump,
- phases are in thermal equilibrium (same temperature).

Internal losses include all hydraulics losses in the fluid mixture (friction losses). This is an approximation, because some heat is evacuated in other ways, for instance through the pump body, lubricant and sealing fluids. The contribution of heat produced by mechanical losses in bearings can be ignored, as it is evacuated primarily by the lubricating system.

Calculation of the temperature rise generally requires use of enthalpy diagrams or of process computer codes performing enthalpy calculations. Calculations based on ideal gas formulas are also useful for preliminary calculations. They are often sufficient, as the temperature rise is low if the compression ratio and GVF are not too high.

If we ignore the change in kinetic energy and elevation between the suction and discharge, the change of enthalpies of the two phases balances the work and heat received respectively:

$$\rho_l q_l dh_l = \delta W_l + \delta Q_l^{loss} + \delta Q_{gl}$$

$$\rho_g q_g dh_g = \delta W_g + \delta Q_g^{loss} - \delta Q_{gl}$$

In these equations, $dh_{l,\,g}$ are the specific enthalpies (enthalpies per unit mass) of liquid (suffix l) and gas (suffix g), $\rho_{l,\,g}$ the phase densities, $q_{l,\,g}$ the volumetric flowrates, $\delta W_{l,\,g}$ the work received by the fluids, $\delta Q_{l,\,g}^{loss}$ the fraction of losses transformed into heat and δQ_{gl} the heat removed from the gas and transferred to the liquid to ensure temperature equilibrium. From the properties of thermodynamic functions:

$$dh = \left(\frac{\partial h}{\partial T}\right)_p dT + \left(\frac{\partial h}{\partial p}\right)_T dp = C_p dT + \left(\frac{1}{\rho} + \frac{T}{\rho^2}\left(\frac{\partial \rho}{\partial T}\right)_p\right) dp$$

or

$$dh = C_p dT + \frac{1}{\rho}\left(1 - \beta T\right) dp \tag{2.13}$$

where C_p is the specific heat capacity at constant pressure and $\beta = -\frac{1}{\rho}\left(\frac{\partial\rho}{\partial T}\right)_p$ the coefficient of isobaric thermal expansion. The relationship can be also written $dh = C_p\left(dT - \mu dp\right)$ where μ is the Joule-Thomson coefficient (isenthalpic expansion).

For an incompressible liquid phase, the first term of the *dp* coefficient balances the compression work. The second one is small and often neglected. This approximation is equivalent to assuming that compression of the liquid does not produce an increase in temperature. In fact, compression of a liquid produces a small increase in temperature, usually negligible when compared with the increase resulting from heat losses or heat transfer. For instance, for water above 40°C: $\beta \approx 4.10^{-4} K^{-1}$ and $\beta T \approx 0.13$. Water compression of 10 MPa produces a temperature increase $\Delta T = \frac{0.13 \times 10.10^6}{1\,000 \times 4\,185} = 0.31K$.

For the liquid phase, we have:

$$\rho_l q_l C_{pl} dT = \delta Q_l^{loss} + \delta Q_{gl}$$

For an ideal gas the coefficient of *dp* is cancelled because the coefficient of isobaric expansion is exactly $\beta = 1/T$. The enthalpy of an ideal gas depends only on its temperature. We have thus:

$$\rho_g q_g C_{pg} dT = \delta W_g + \delta Q_g^{loss} - \delta Q_{gl}$$

and finally for the mixture:

$$\left(\rho_l q_l C_{pl} + \rho_g q_g C_{pg}\right) dT = \delta W_g + \delta Q_{l+g}^{loss} \tag{2.14}$$

The heat generated by losses δQ_{l+g}^{loss} can be related to the internal efficiency of the pump η_i. According to the definition of this efficiency, the required mechanical work is $\delta W_t = \delta W_i / \eta_i$, and the losses are $\delta Q_{l+g}^{loss} = \delta W_t - \delta W_i = \delta W_i\left(1 - \eta_i\right)/\eta_i$. After substitution in the above equation:

$$\left(\rho_l q_l C_{pl} + \rho_g q_g C_{pg}\right) dT = \frac{\left(1 - \eta_i\right)}{\eta_i}\delta W_l + \frac{1}{\eta_i}\delta W_g$$

The non-symmetric form of the right-hand side is the consequence of the behaviour difference between the liquid and gas phases. Using the ideal gas relationship, this equation can be transformed to:

$$\left(\rho_l q_l C_{pl} + \rho_g q_g C_{pg}\right)\frac{dT}{T} = \frac{1}{\eta_i}\rho_g q_g\left(C_{pg} - C_{vg}\right)\frac{dp}{p} + \frac{\left(1 - \eta_i\right)}{\eta_i}\rho_l q_l \frac{dp}{T\rho_l}$$

or, using the gas mass fraction $x_g = \frac{\rho_g q_g}{\rho_l q_l + \rho_g q_g}$:

$$\frac{dT}{T} = \frac{1}{\eta_i}\frac{x_g\left(C_{pg} - C_{vg}\right)}{\left(\left(1 - x_g\right)C_{pl} + x_g C_{pg}\right)}\frac{dp}{p} + \frac{\left(1 - \eta_i\right)}{\eta_i}\frac{\left(1 - x_g\right)}{\left(\left(1 - x_g\right)C_{pl} + x_g C_{pg}\right)}\frac{dp}{T\rho_l} \quad (2.15)$$

Let us now define an ideal polytropic coefficient k for the mixture:

$$\frac{k-1}{k} = \frac{x_g\left(C_{pg} - C_{vg}\right)}{\left(\left(1 - x_g\right)C_{pl} + x_g C_{pg}\right)}$$

The value of k is thus given by:

$$k = \frac{\left(1 - x_g\right)C_{pl} + x_g C_{pg}}{\left(1 - x_g\right)C_{pl} + x_g C_{vg}} \quad (2.16)$$

For the mixture, this coefficient is equivalent to the gas ratio C_{pg}/C_{vg} and it is effectively the ratio of the phase specific heat capacities at constant pressure and at constant volume since the assumption of an incompressible liquid leads to $C_{vl} = C_{pl}$. It is also consistent with the conventional relationship giving the polytropic coefficient n of the actual compression used in compressor studies:

$$\frac{n-1}{n} = \frac{1}{\eta_i}\frac{k-1}{k}$$

Integration of Equation (2.16) between two states from $\left(p_1, T_1\right)$ to $\left(p_2, T_2\right)$ gives:

$$T_2 = T_1\left(\frac{p_2}{p_1}\right)^{\frac{n-1}{n}} + \frac{\left(1 - \eta_i\right)}{\eta_i}\frac{\left(1 - x_g\right)}{\left(\left(1 - x_g\right)C_{pl} + x_g C_{pg}\right)}\frac{p_2 - p_1}{\rho_l} \quad (2.17)$$

When the mixture has a very low gas content $\left(x_g \to 0\right)$, the fluid evolution is in practice isothermal $\left(k \to 1\right)$. The temperature rise results only from heating by the losses (second term).

When the mixture is in practice comprised only of gas $\left(x_g \to 1\right)$, $k \to \frac{C_{pg}}{C_{vg}}$ and the usual formulas for compressors are obtained with $T_2 = T_1\left(\frac{p_2}{p_1}\right)^{\frac{n-1}{n}}$.

2.5.3.3 Efficiency

Energy losses are not governed by the same phenomena in twin-screw pumps and helico-axial pumps. Due to these differences, it is difficult to establish a fair comparison of the two concepts based on the energy consumption criteria. It is often possible to find a situation where one concept is better-suited, in term of efficiency, than the other. The actual operating point of the pump, compared to the nominal best efficiency point (BEP), also has a considerable influence on the effective efficiency of the pump for both types.

In twin-screw pumps, the mechanical energy is mainly transferred to the multiphase mixture by an isochoric process (*i.e.* at constant volume). The compression work received by the fluids follows a near-isothermal process. The difference between the two energies, represented by the two areas in Figure 2.18 (PV diagram) is lost by backflows and pressure drops clearances to compress the gas. The corresponding efficiency depends on the pressure ratio. It drops from some 70% at a compression ratio of 2, to some 37% at a ratio of 6, and even less at higher values. Additional efficiency losses are produced by the volumetric efficiency and fluid friction in screw clearances.

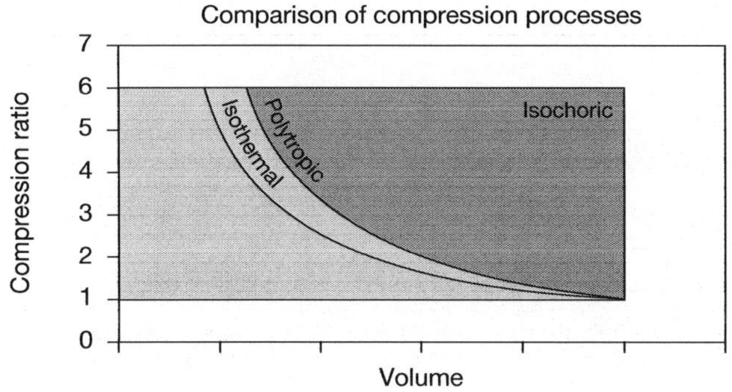

Figure 2.18

Compression work areas in a twin-screw pump

Helico-axial pumps show relatively large friction losses because of the shape of the compression cells having large wetted areas. The basic efficiency is about 65% – 55% in single-phase flow for this type of pump, depending on the pump size. In two-phase flow, additional efficiency losses are caused by the internal slip of velocities and non-equilibrium of temperature between the phases. They are primarily governed by the mixture GVF and the ratio of phase densities.

MPPs of all types have relatively low energy efficiency when compared with modern single-phase pumps or compressors. As already mentioned, a part of this low efficiency is not related to the quality of the machine but comes from simplifications in evaluating the compression energy of the multiphase mixture, for example by assuming it is a homogeneous mixture without velocity slip at thermal equilibrium and by generally assuming the compression to be isothermal. The influence of the isothermal assumption on efficiency is well-known to compressor users. The isothermal efficiency is always lower than the adiabatic or polytropic efficiencies.

The best way to assess multiphase pumping in terms of efficiency, *i.e.* in terms of energy, is probably to compare the necessary compression energy to the energy consumption that alternative methods of production would require, for instance separation with pumping of liquid and compression of gas, or the use of gas-lift in wells. Multiphase boosting is generally a better solution (in terms of energy efficiency) than gas-lift.

2.5.4 Transient Behaviour

In actual field conditions, an MPP must be able to handle several types of unsteady or transient flows in operating conditions. In these conditions, the pump must not impair the stability of flow.

Good knowledge of transient behaviour is required when establishing operational procedures such as shutdown or restart, but also when selecting and designing the control system. Transient behaviour depends on the complete production system and is influenced by the control system.

The usual approach is to study the response of the system to sudden changes in flow parameters such as flowrates, gas fraction, pressure, etc. A more sophisticated way is to simulate performance in slug flow and with certain specific process trips, incidents or failures, such as a sudden shutdown of a well, closure of a downstream valve, or a sudden loss of drive power.

Such studies are made easier by the availability of a multiphase pump performance model coupled to a dynamic process simulation software or a transient pipeline multiphase flow simulation code.

To study the transient behaviour of a helico-axial pump, Ramberg & Bakken [1997] integrated a simplified model of a helico-axial multiphase pump into a process simulator. The multiphase head is described as a degradation of the single-phase liquid head, influenced by the GVF and the density ratio of the gas and liquid phases. A simple but approximate formula is used. The single-phase liquid head is given by a linear function of the total volumetric flowrate. An isothermal flash provides the GVF and phase densities in suction conditions (pressure and temperature), an isentropic flash then is performed to obtain the temperature and GVF at discharge.

With this tool, the authors studied the transient behaviour of an MPP operating on the Gullfaks platform (North Sea) subjected to a slug flow, to a sudden well shutdown or to closure of a downstream valve. Although simplified, the model was able to simulate the actual transient behaviour of the MPP with an acceptable degree of accuracy within the tested range.

In another approach, a full multiphase pump model, adapted to the helico-axial concept, was integrated in the Tacite multiphase pipeline simulation code to study transient flows in a pipeline [Heintzé *et al.*, 1999]. The Tacite code is based on a compositional description of the fluids and can simulate complex thermodynamic behaviour when the temperature and pressure in the pipeline change. A multiphase pump module is integrated into the code. At each time step in a simulation, the pressure rise output by the pump is calculated from the pump characteristics (number of stages, stage characteristics, pump rotational speed) and the process conditions (GVF, phase densities, flowrates, etc.).

The simulation tool may be used to study transient behaviour during normal operating procedures such as production shutdowns or restarts, but also to simulate various types of incidents. For instance, Figure 2.19 displays the path followed by the operating point in a "head versus flowrate" diagram, during a planned shutdown followed by a restart. The word "planned" means that the operations are anticipated and that the pump rotation speed is adjusted in steps to the wells shut-in time to obtain a smooth system response. The diagram shows that the operating point remains centred in the pump domain at any phase of the shut-down and restart procedures.

Such simulations are an effective means of ensuring the safety and reliability of the production system with a multiphase pump, especially for unmanned fields or subsea production.

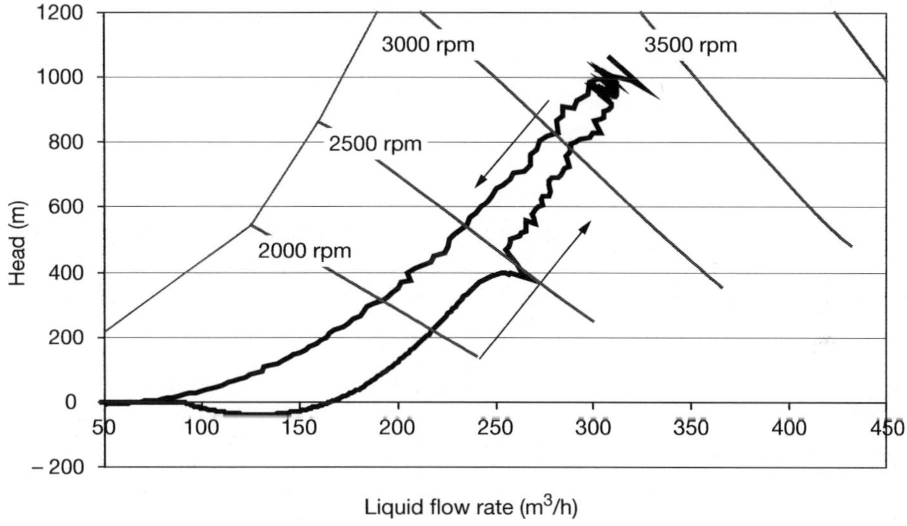

Figure 2.19

Evolution of MPP operating point during a planned shut-down and restart

2.5.5 Pump Control

Different levels of pump control can be used, depending on the operational requirements. Standard practice calls for the appropriate level of mechanical protection for the pump and driver units. Alarm values are typically used to prevent damage to the equipment.

2.5.5.1 Flow Homogeniser

In case of strong incoming slug flow with a succession of liquid slugs and gas pockets at a period of less than one or two minutes, the sudden changes in mixture density or composition produce associated variations of drive power and rapid and repetitive torque changes or pulsing. This behaviour can be eliminated simply with a flow homogeniser (Figure 2.20, top diagram).

The homogeniser is basically a small vessel with a static mixer that is installed upstream of the pump. Its main functions are to dampen the energy of liquid slugs and to smooth out GVF fluctuations. It can also act as a sand trap.

The flow homogeniser can be also useful at start-up of a well to avoid dry running of the pump during production of the gas pocket accumulated in the upper part of the well (Figure 2.20, bottom diagram).

Obviously, when the flow characteristics only vary slightly from the average values, there is no need to install a homogenizer.

It has been already mentioned that the bottom casing of twin-screw pumps acts as a homogeniser. It is often unnecessary to add an additional capacity for this type of pump if long dry running periods are not anticipated. For down-hole rotodynamic pumps, the well itself acts as a homogeniser.

2.5.5.2 Self-Adaptability Capability

The self-adaptability capability of rotodynamic MPPs can be illustrated with the specific example of slug flow. When the instantaneous GVF of the mixture entering the pump rises, the differential pressure generated by the pump diminishes (Figure 2.21). The discharge pressure stays practically constant, at least initially, because this pressure is governed by the separator pressure plus the pipeline pressure losses, which are not disturbed as long as the flow has not changed significantly. The system response is an immediate increase in pressure at suction, producing a reduction in the total flowrate and therefore an increase in the differential pressure. At the same time, the mixture density rises, contributing to the feedback effect on the differential pressure. The reverse mechanism also takes place when the GVF decreases.

This behaviour has actually been observed both on multiphase flow loops and on production sites [Ramberg & Bakken, 1997].

2.5.5.3 Process Control

Well flowrates at a given time depend essentially on the wellhead flowing pressure (WHFP). As the pump discharge pressure is appreciably constant, as explained above, flow fluctuations mainly involve suction pressure fluctuations which in turn will affect well production.

In normal conditions, a constant oil flowrate is generally required with a pump operating at constant speed. To keep well output constant, the pump rotation speed can be controlled. A speed reduction (or rise), decreases (or increases) the pump differential pressure. The suction pressure will therefore rise (or fall) to balance the influence of the flow fluctuations.

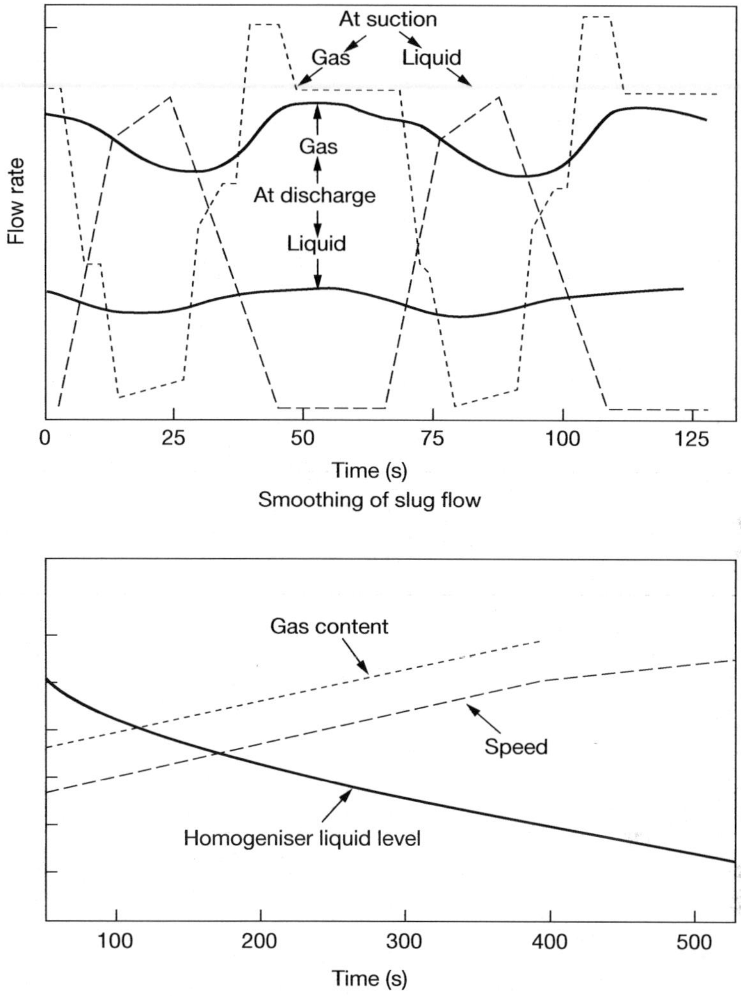

Figure 2.20

Flow homogeniser behaviour

However, fixed speed installations are possible as long as one can tolerate a certain amount of back-pressure fluctuation in the wells (or the upstream pipeline, if any) and in field production. For more critical applications, variable speed drive allows full process control.

A first level of process control can be achieved by manual adjustment of the pump speed. The field operator can, every day or every week, set a predetermined speed to produce the required output.

A manual control system can still offer the required mechanical protections, but it only allows manual speed setting. The system will not by itself be able to adjust the speed auto-

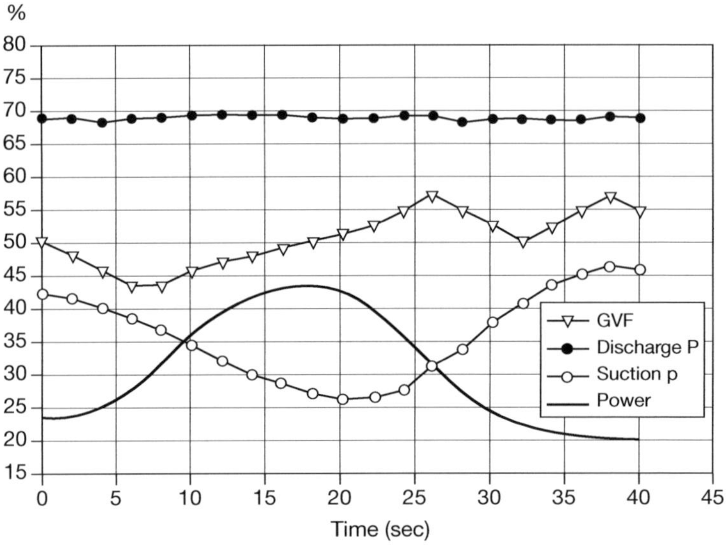

Figure 2.21

Pump stability (after Ramberg & Bakken, [1997])

matically. The pump will either run at the pre-determined speed or be shut-down. Obviously, the speed setting can be changed manually by the operator at any time. Such control would also enable adjustments to be made over a period of time to balance out the natural depletion of the field.

More elaborate process control can also be achieved through automatic setting of pump speed. This allows an operator to set and maintain a required production output by avoiding influences from upstream or downstream disturbances, such as changes in settings of manifold valves, chokes, well control, and by preventing unnecessary shut-downs.

In some cases, fixed-speed operation provides enough operational flexibility. In other cases, where back-pressure control on the wells is more critical or where adaptability to evolution of operational parameters is essential, a variable speed system is preferable. Pump control and monitoring can be more or less sophisticated depending on the application, expected well behaviour, pump redundancy, importance of maintaining a continuous production flow, manned or unmanned facilities.

2.5.6 Monitoring

As with other pumping units used in oil production, multiphase pumps are typically equipped with conventional instrumentation suitable for outdoor operation and hazardous areas:

- pressure and temperature sensors at suction and discharge;

- pressure and temperature sensors on oil lubrication and sealing circuit;
- bearing temperature monitors;
- rotation speed measurement (in the case of variable speed drive).

2.6 FIELD APPLICATIONS OF HELICO-AXIAL PUMPS

This section reviews field deployment of helico-axial multiphase pumps with an overview of few selected application cases. Data obtained from field experience with these pumps are summarised in order to give information on the characteristics of the most relevant applications and on the benefits obtained from multiphase pumping in the hydrocarbon production industry.

2.6.1 The Early Stage: Field Demonstrations

The first industrial prototype of helico-axial multiphase pumps was tested by Total in a Tunisian field in 1991 within the framework of the Poseidon joint R&D project. After the success of the experiments, the multiphase pump technology was transferred to two licensed manufacturers: Sulzer Pumps and Framo Engineering, and two additional machines were built by the manufacturers. Since then, several of these pumps have been manufactured for various field applications.

These two prototypes were horizontal axis pumps designed for surface applications, one onshore for Elf on the Pecorade field (France) [Falcimaigne *et al.*, 1994; Leporcher et Taiani, 1995] and a second offshore for Statoil on the Gullfaks A platform (Norwegian North Sea) [Vangen *et al.*, 1995]. Both had an electrical asynchronous drive with variable speed between 3,000 and 7,000 rpm and nominal power in the 600 to 750 kW range, corresponding to a hydraulic power of approximately 500 kW.

The applications of multiphase boosting that were clearly apparent to the oil fields operators at that time were:

- restoring production of closed wells,
- marketing gas from low-pressure wells instead of flaring.

The Pecorade demonstration test case corresponded to a declining oil field application with low wellhead flowing pressure whereas the Gullfaks application, with approximately 60 bar flowing pressure at pump suction on the platform, was more to meet the need to boost a well stream from a subsea cluster up to a given pressure in order to restore the production of closed wells in an existing production facility.

From the production engineering point of view, the lessons from the two demonstration tests in actual production conditions were mainly that:

- Working with a fluid flow at constant GVF, rotodynamic multiphase pumps behave like positive-displacement pumps at high compression ratio with the steep characteristic curves of the helico-axial pump at constant rotation speed. An example of the per-

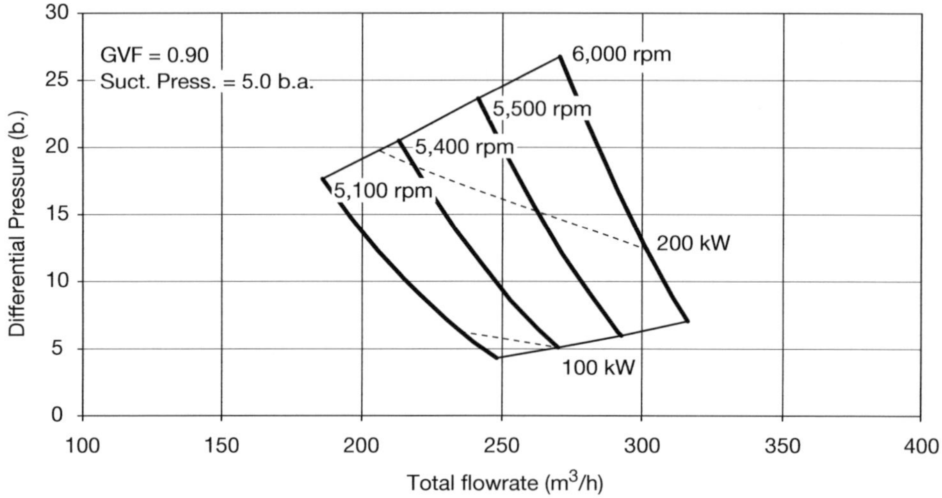

Figure 2.22

Example of multiphase pump performance curves (P_{suc} = 5.0 bar, GVF_{suc} = 0.90) – see also Leporcher & Taiani [1995]

formance curves of the pump installed on the Pecorade field is shown at Figure 2.22 in order to illustrate this.

- Equipped with a flow homogeniser at suction, the pumps can be run effectively at a constant speed of rotation in stabilised production conditions thanks to the self-regulating capability of the rotodynamic hydraulic concept.
- When used on a single well stream, the pressure difference given by the pump primarily has the effect of lowering the wellhead flowing pressure, while the discharge pressure is usually kept constant by the pressure-controlled downstream processing facilities. This can be used profitably because the decrease in the wellhead pressure is generally accompanied by a substantial increase in the flowrates of produced gas and oil.
- In transient production conditions or when the wellhead pressure is modified significantly due to the pump pressure difference, the gas content and the flowrate variations depend on the specific case. Therefore, the operation and control of the multiphase pumps must be adapted according to the specificity of the production case and the objectives of the operator.

As this information was analysed, the operators rapidly became aware that some unexpected benefits could be gained through the use of multiphase boosting. In fact, in addition to the applications initially envisaged, early field experience indicated that multiphase pumps could be also considered for,

- increasing the oil production rate by reducing the wellhead back-pressure,
- use as an additional artificial lift technique, alongside the gas-lift technique, for multiphase field developments.

This opened up a wide range of possible applications for multiphase pumps in the oil industry, based on the fact that they can be used as a new type of production management tool which allows “positive” choking of the produced stream.

2.6.2 Overview of Typical Surface Applications

2.6.2.1 Samotlor – Western Siberia

Two pumps were manufactured by Sulzer Pumps and installed in 1997 on the giant Samotlor field (light oil, with sand production) operated at that time by Chernogorneft [Korolev *et al.*, 1999]. The production is collected from various clusters (of about 50 wells) located at distances up to 5.5 km from the multiphase pumps station and the pumps are used to boost the pressure in order to transport the effluent to the central processing station located 15 km away.

The two pumps run in parallel with a total capacity of approximately 1,000 m^3 per hour. The suction pressure fluctuates while the GVF in suction conditions varies between 70 and 86%. The control system keeps the suction pressure within a chosen interval. The discharge pressure is up to about 20 bar. The outside temperature varies between – 40°C and + 35°C. Each pump is driven by a 400 kW fixed-speed electric motor and is capable of running from 1,500 to 4,000 rpm through a step-up gearbox and hydraulic coupling allowing adjustment to take account of production requirements and field decline. The multiphase booster units, with drive train and lubrication system, are mounted horizontally on a common skid inside an insulated shelter equipped with heating and ventilation because of the Siberian climate. Overall dimensions are 8 m long by 3 m wide by 3.5 m high. Total weight is 18 tons, shelter included. The control panel and electrical utilities are installed at some distance in a separate shelter in a non-hazardous area.

This case presented some interesting features:

- experience of two pump running in parallel,
- severe slugging conditions,
- high sand content,
- very harsh environmental conditions.

All these difficulties were successfully surmounted. Since the first installation, a number of other helico-axial MPP of higher capacity have been installed in the field (with driver power up to 2,000 kW).

2.6.2.2 Duri – Indonesia

One horizontal pump was installed in 1998 on the Duri field (Indonesia) operated by Caltex. The pump was used to transport production from a satellite to the production centre along a 10 km pipeline. Duri is a viscous oil field and production is assisted by steam injection. The multiphase pump collects the effluent, which includes hydrocarbon liquid and vapour, water in liquid and vapour phases, a high concentration of CO_2 and a high content of sand.

The characteristics of this application case are:

- Total flowrate: 1,000 m^3/h.
- Inlet GVF: 40 to 75%.
- Inlet pressure: 6 to 13 bar abs.
- Discharge pressure: up to 18 bar abs.
- Drive power: 550 kW.

This case presented specific difficulties:

- high temperature,
- condensation of steam,
- very high sand content.

Due to the difficult process conditions and the considerable changes in flow characteristics (steam condensation) the pump was stopped after several months of operation on the field.

2.6.2.3 Dunbar – Offshore North Sea

Two very high capacity Poseidon pumps manufactured by Sulzer Pump were installed in 1999 and have operated since then on the Dunbar platform operated by Total. These two pumps, running in parallel, boost the pressure of the low-pressure wells in the Alwyn South area to reach the pressure of the pipeline which exports the production from Dunbar to the Alwyn North Platform. The design characteristics of each pump are:

- Total flowrate: 1,200 m^3/h.
- Inlet GVF: 30 to 90%.
- Inlet pressure: 50 to 70 bar g.
- Discharge pressure: 125 bar g.
- Drive power: 4,500 kW.

Multiphase boosting was preferred to the other options carefully studied by Total for phase II of the Dunbar field development [Delaytermoz, 2000; Leporcher *et al.*, 2001]. After the initial period of natural depletion, which began in 1994, the field needed additional equipment on the Dunbar platform in order to maintain production. The conventional revamping option originally scheduled, consisted mainly in the installation of a new HP separator on the Dunbar platform and the laying of a new 16-inch pipeline to export the HP gas phase to the Alwyn North central processing facility. The HP liquids would have been mixed with the LP production stream to be exported *via* the existing 16-inch multiphase pipeline operated at a lower pressure of 30 bar.

The study of the conventional option showed that it also implied extensive debottlenecking modifications on the LP system in the Alwyn North facility and that a risk of flow instabilities (slugging) existed if the multiphase pipeline were to be operated at low pressure. On the other hand, the multiphase boosting option enabled the LP wells to be boosted directly into the existing pipeline. Consequently, construction of a new 25 km pipeline was not necessary. The multiphase boosting option offered a flexible solution to export the increasing LP production from the Dunbar platform at an initial investment cost less than half the cost of the conventional upgrade. The main disadvantages of the multiphase option are the additional weight on the platform, 700 tons instead of 450 tons, and the level of confidence that decision-makers could have in such a new technology.

The multiphase pumping station was operated at two different suction pressures with a strategy of routing the wells through the pumps, enabling hydrocarbon export through the existing production system to be maximised. In the future, depending on well behaviour, a third multiphase pump will be installed for operation in series with one existing unit, thus enabling the well head flowing pressure to be lowered again, from 50 bar to 20 bar.

At the 2005 Multiphase Pump User Roundtable, Bibet [2005] presented production trends at Dunbar over the 5 past years and the consequences on the pumping operation. During this period, the gas-liquid ratio at the pump inlet increased from 3.5 to 10 (*i.e.* the GVF increased from 77% to 91%). At the same time, the inlet pressure was reduced from 70 bar to some 40 bar with less stable well production. Some modifications were made to the installation and the pumps to avoid the vibrations and process trips caused by flow instabilities. In particular, the outlet device of the flow homogeniser was redesigned to operate in the correct GVF range, the pump throttle bushings were also redesigned to increase the damping effect and a power control was used instead of the conventional speed control.

To conclude, helico-axial MPP appeared to be a flexible solution, able to handle large process variations with minor changes. The case analysis also showed the importance of having a homogeniser well-suited to the actual flow characteristics.

2.6.2.4 Lennox – Offshore Irish Sea

A 1 MW, electrically-driven multiphase pump manufactured by Framo Engineering has been successfully used since August 2003 on the unmanned Lennox platform operated by BHP Billiton Petroleum in Irish Sea. In this application. The pump is mainly used for supplementing artificial lift and well kick-off after shut-in periods [Donachie, 2004].

Difficult wells on this declining field used to be restarted on an intervention basis before the installation of the multiphase pump. This resulted in production losses of up to 10000 bbl/day over the periods of several days needed to mobilise the intervention crew in order to kick-off shut-in wells. In addition, with GVF increasing due to reservoir depletion it was anticipated that many wells would require permanent artificial lift by the year 2005.

The decision to opt for multiphase boosting was taken after a comparison with the other possible technologies. Gas lift was not selected because of the need for facility upgrade, including compression and process system. This option implied considerable expense and several workovers with production interruption to install gas lift completions. Electrical Submersible Pumps (ESP) were not a viable artificial lift retrofit technique for this application due to the high GOR coupled with production below the bubble-point pressure. The multiphase pump therefore offers an easy solution for quick restart of shut-in wells with no intervention required on the unmanned platform and great operational flexibility.

When necessary, the pump will be used on a continuous basis to provide supplemental lift to the lower pressure wells thus helping to maintain production.

2.6.2.5 Priobskoye – Western Siberia

The gradual accumulation of field experience gave the oil operators the confidence they needed to apply the technology on a larger scale. The Dunbar case, presented above, was a significant step in this process. The Priobskoye case was another milestone. The multiphase pumps installed in this field, manufactured by Sulzer Pumps, are the largest in service today.

The pumps are installed in the part of the Priobskoye field extending on the right bank of the River Ob, in Western Siberia. Production initially started 10 years ago on the left bank, where the separation and treatment facilities are located. On the right bank approximately 24,000 bopd were produced in 2000 with significant increase potential. This part of the field is operated by the JSC Yujanskneftegas operating company. The oil was transported along a 33 km long pipeline, 426 mm in diameter, to the treatment facilities on the left bank.

The climate is harsh (ambient temperatures range from – 55°C to + 35°C). The area is swampy and difficult to access when the River Ob floods in spring and summer. This is an also environmentally sensitive and protected area.

A second pipeline has been built to bring total production to 200,000 bopd. Two concepts were considered for this development, multiphase pumps and a conventional separation station. The MPP alternative was selected because it does not require permanent manning, flaring, or venting to atmosphere, or discharge in water, as well as a significant cost saving [Pershukov *et al.*, 2001].

Two pumps were commissioned and started in summer 2001 and a further two in 2002. Each pump is designed for a total suction flowrate of 3,300 m^3/h and they are driven by a 6 MW variable-speed electric motor. Owing to the severe environment, the pumps are installed in a weatherproof shelter along with the electric motor, gearbox and ancillary seal and lube oil systems, suction and discharge valves and non-return valve. A second shelter for each pump contains the variable-frequency drive, the motor control cabinet, instrumentation panel and a local control room. The shelters are completed with heating, ventilation and fire and gas detection systems. Equipment common to all 4 pumps include a single flow homogeniser and an electric distribution centre which houses a 35 kV switchboard and 2 off 400 volt transformers.

The pumps are designed to operate in automatic mode and are controlled from a remote operation centre due to the access difficulties in spring and summer. The pumps are operating initially with a moderate differential pressure and are de-staged. Additional stages will be added later. The use of different cartridge designs enables the pump duty to be adapted to suit changing field requirements.

2.6.3 Subsea Pumps Development and Applications

Besides the field deployment of surface MPPs, subsea machines were seen very early on as a promising alternative for the production of marginal oilfields [Arnaudeau, 1988]. This particularly applies in deepwater conditions where the pump could provide the pressure increase necessary to compensate for the hydrostatic gradient in the riser.

For subsea applications, the satellite multiphase booster offers certain advantages when compared to downhole electro submersible pumps:

- being installed at the wellhead, it is more accessible than the downhole pump which requires well entry and it is therefore cheaper to maintain,
- a single pump unit can be used to boost several wells.

2.6.3.1 Draugen – North Sea

An experimental subsea boosting station, the first in operation in the world, was installed in 1993 on the southern water injection template of the Draugen field operated by Norske Shell on the Norwegian side of the North Sea [Darrigan *et al.*, 1995]. Started in 1994, the subsea station delivered by Framo Engineering, named SMUBS (Shell Multiphase Underwater Booster Station), boosted the effluent coming from the Rogn South subsea well. There are 4 km of subsea flow line transporting the effluent free flow from the Rogn South wellhead to the southern template. The effluent was then boosted by SMUBS and transported to the Draugen platform through a 6 km subsea flow line. This pressure boosting allowed production of an additional 40%, or 600 m^3/d of oil. SMUBS was driven by a hydraulic turbine activated by the water injected in the field. However, over longer distances, the hydraulic losses and the cost of the line providing the high pressure water drive become significant. Electric power is therefore a more practical source of energy for most satellite applications.

After the full-scale tests of other concepts [Cordner & de Salis, 1995; Wysatta, 1996], the above designs have been chosen for all the subsea multiphase boosting facilities running to date.

Their configuration was developed to meet the requirements of subsea installation and operation procedures, based on the philosophy of keeping the system and its components as simple as possible. All components that are susceptible to wear are located in a single retrievable cartridge for easy installation and maintenance, with a single ROV simply for monitoring.

Therefore, the characteristics of the pumps for subsea applications are mainly:

- rugged system, minimising complexity,
- high reliability,
- packaged system for ease of installation, maintenance and relocation.

2.6.3.2 Topacio – Subsea Gulf of Guinea

Two 1 MW, electrically driven, subsea Poseidon pumps were installed in 2000 on Exxon-Mobil's Topacio satellite field in the Gulf of Guinea. The field is located in 550 m of water and is tied back to an FPSO moored on the neighbouring Zafiro field, 8.5 km away. The design characteristics of each pump are:

- Total flowrate: 470 m^3/h.
- Inlet GVF: 75%.
- Inlet pressure: 15 bar g.
- Discharge pressure: 50 bar g.

- Drive power: 840 KW.
- Design rotation speed: 5,060 rpm.

The electric power to the pump is supplied through a dedicated cable/umbilical laid from the FPSO and the pumps control panel is located on the FPSO. A subsea transformer module is installed close to the pumps module to deliver the required voltage to the electric motors.

The multiphase pumps at Topacio essentially apply a significant drawdown on the four boosted wells, resulting in increased production, thus enabling the operator to maintain the target production from the total Zafiro field. But, in addition, the increased flowrates obtained with multiphase boosting also resulted in mitigation of slugging in the flow line, leading to more stable production with an increased process capacity and reduced pressure losses in the pipeline.

2.6.3.3 Ceiba – Subsea Gulf of Guinea

The Ceiba field came on stream in November 2000 with production from 4 early production wells. The management's objectives for this project lead to the use of multiphase boosting technology as an early production system prior to fast-track full field development [Garrett & Drinkard, 2002].

Four 1 MW subsea multiphase pumps manufactured by Frano Engineering were installed in 750 metres of water at the heads of the first four producing wells drilled on the Ceiba field. The need for artificial lift was identified early on in the field development project and the pumps were started up only a few months after drilling and used to tie in the wells with the FPSO anchored at a distance of 7.5 km. The pumps work independently on the well streams of the four wells and boost the total production into 8-inch steel flow lines.

The design characteristics of each pump are:

- Total flowrate: 250 m^3/h.
- Inlet GVF: 50 to 90%.
- Inlet pressure: 16 to 50 bar abs.
- Discharge pressure: 50 to 70 bar abs.
- Drive power: 860 KW.

As for the Topacio application, electric power is supplied to the pump through a dedicated cable/umbilical, plus a subsea transformer module, laid from the FPSO. The pumps control building is located on the FPSO.

It should be noted that, for the Ceiba project, the pump manufacturer was able to deliver an integrated early production system for a subsea application since the scope of supply included the multiphase booster pumps, their subsea support structures complete with valves, the subsea transformers, the power/hydraulic/signal cable-umbilical, the topside pump control building and the pump intervention system.

This approach allowed production to be started from low-pressure wells with good control of FWHP and slugging in the flow line within just two years of field discovery.

Several other subsea multiphase boosting systems have been delivered since these application cases.

2.6.4 Downhole Applications

One advantage of Electrical Submersible Pumps (ESP) over the other artificial lift techniques has been the ability to draw down the wells to the minimum flowing pressure to produce high rates. This technique, however, has been limited by the inability of the centrifugal impellers used in ESP to handle free gas at their intake. Consequently, ESP have been limited to wells producing no free gas, or are sometimes, used in conjunction with gas separation devices.

The rotodynamic helico-axial multiphase pumps concept, adapted to impeller sizes for downhole installation, was developed in the late 90's with the objective of increasing the gas handling capability of ESP. In this application, the multiphase compression cells are installed in series in the upstream part of the ESP, in sufficient numbers to lower the GVF at the intake of the first centrifugal impeller to a proportion of a few percent that can be handled efficiently by a standard centrifugal ESP. Thus in down-hole industrial applications, this particular design can be used as a gas handler connected below the conventional ESP. The ability to generate a pressure increase at high GVFs will enable the stages of the submersible pump to handle gas better and prevent gas-locking.

A commercial application of this type took place in the San Francisco field in Colombia where an ESP equipped with a Poseidon type multiphase pump/gas handler at the intake was installed in the San Francisco-75 well, in 2002 [Villamizar, 2003]. The field, operated by Hocol SA, has been in production since 1984 *via* more than 100 wells, each with up to 12 different producing layers in the formation from which the fluids are commingled using submersible pumps.

In 2000, a gas injection project was started in the field as an alternative recovery technique to improve sweep once the water cut in the production reached 90%. Gas injection effectively increased oil recovery, although one of the drawbacks was the breakthrough of high volumes of gas at the producer wells.

A rapid increase in the volume of gas produced was then observed in SF-75 and in the other producing wells. As a consequence, oil production from SF-75 was limited by the amount of gas that the submersible pump was capable of handling. In addition, the well was shutting down due to gas-locking and motor current was unstable owing to the presence of gas in the pump. This resulted in significant production losses and convinced the Hocol management to replace the existing ESP by a new string comprising of a submersible pump plus a Poseidon type "gas handler" plus a VGS gas separator at intake.

The multiphase ESP installation rapidly proved to be a good solution for the SF-75 gas problems. In comparison with the conventional ESP performance before replacement:

- liquid gross production rate was increased by 40%,
- oil production rate was increased by 100%,
- more gas could be produced through the pump during the gas injection cycles,
- the pump handled more than 60% free gas at intake,
- gas locking shutdowns and current fluctuations were eliminated.

After the success of this industry first, Hocol is considering expanding the gas injection project to cover more wells in the San Francisco field, relying on the performance of the multiphase ESP to produce wells with high GVF.

2.7 CONCLUSION

Multiphase production which leads to simpler and smaller facilities for hydrocarbons production is now made possible thanks to the availability of a full range of multiphase pumps. Extensive research and testing on actual production fields has been carried out to develop the technology. Excellent operational results have been obtained, confirming the soundness and versatility of the technology over a wide range of operating conditions.

In previous years, multiphase pumps have proven to be a cost-effective means of enhancing production. Furthermore, with growing field experience, the operators are now confident that multiphase boosting can be regarded as an effective production management tool.

In deepwater and mature areas, multiphase boosting offers a solution for enhancing production, increasing total oil recovery and, in some cases, helping to solve problems such as slugging or gas hydrates formation when installed close to the well head [Leporcher *et al.*, 1999].

For new development projects, subsea multiphase pumps used as an early production system can reduce the time lag between reservoir discovery and initial oil production. They could also help put off and facilitate the decision to opt for full field development. The development of long tie-back satellite fields will be made easier with the availability of MPP.

Helico-axial pumps cover a broader range of applications than expected in the early stages of industrial development. They can be used efficiently with very high gas volume fractions and low suction pressures. Their inherent low weight makes them particularly attractive for large flowrates and offshore operations. The continuously-open channel of the pumps is also an advantage when solid particles are produced.

Chapter 3

Multiphase Metering

3.1 INTRODUCTION

In petroleum production, well streams are characterized by a mixture of fluids, oil, gas, water, often laden with solid particles (sand, scales, etc.). Until recently, crude oil production was only metered after separation and treatment of the different phases, using conventional single-phase flow instrumentation such as turbine meters, orifice plates, etc. To meter each well separately, a dedicated test separator was required, either mobile or installed permanently on the field.

The need for directly metering well stream production became acute in the 80's with the multiplication of subsea satellite developments and unmanned fields in remote onshore areas. The construction of a dedicated flow line to test wells at a distance is expensive. It also takes a long time (several hours) to purge the line and stabilise the flow between each test. This can create technical problems which could impair test significance, such as severe slug flow. At the same time, advances in reservoir management and modelling have increased the need for frequent well tests to provide data.

Technical and economic needs, in association with a huge potential market, drove extensive R&D efforts. A number of approaches have been explored to select suitable measurement methods [Lynch, 1991]. After a decade of developments and tests, the first meters were marketed in the second half of the 90's. A second generation of meters is today generally proposed by vendors and the number of installed devices is growing steadily [Falcone *et al.*, 2002].

Multiphase flow metering is not simple and a number of complex problems had to be solved. For instance, a basic principle of metrology requires stabilising the flow to obtain a steady velocity profile in the measurement section. Multiphase flows are intrinsically unstable, with several different flow patterns, continuously fluctuating phase fractions and often differing velocities for the same phase (according to the size of bubbles for instance in dispersed flows). In addition, specific problems related to the crude oil composition, such as wax or scale depositions, can disturb the measurements.

At present, the accuracy of multiphase meters is in general better than ± 10% of the measured flowrates in their normal operating domain and exceeds ± 5% for the best meters. Although this accuracy may appear relatively low in terms of metrology performance, multiphase meters would now appear to be as accurate as conventional test separators for testing wells. Their accuracy is also considered to be acceptable for reservoir management and production purposes. Their main uses are monitoring of production trends, identification of problems (water breakthrough, scaling, wax, etc.) and allocation of production between wells or reservoirs when fluids from different leases are commingled. However, multiphase meters are not suitable for fiscal metering which requires accuracy of between ± 0.1% and ± 0.5%. Further R&D work is still needed to achieve this degree of precision.

Multiphase flow meters require less space than test separators and are also easier to transport from field to field. They can also deliver real-time monitoring of the field production. For all these reasons, they are being increasingly used not only subsea but also on topsides or onshore in place of conventional test separators. Very early on, the oil operators recognized the benefits of multiphase metering [Smith, 1995; Shen, 1997]. New applications are now looking at wet gas metering and down-hole metering.

The growth of multiphase metering has in recent years been accompanied by the issue of guidelines and recommended practices which to some extent make up for the lack of established standards. For readers who wish to examine the subject more closely, the most significant publications are the *Guideline Notes for Petroleum Measurements under the Petroleum (Production) Regulations* by the UK Department of Trade and Industry [2003], the fully-rewritten version of the *Handbook of Multiphase Metering* by the Norwegian Society for Oil and Gas Measurement [Dahl *et al.*, 2005] first issued in1995 [Dykesteen *et al.*, 1995], and the *Recommended Practice 86 for Measurement of Multiphase Flow* from the American Petroleum Institute [2005]. Several other publications of interest can be also looked at, such as the draft version of ISO TC 193 *Allocation of Gas and Condensate in the Upstream Area* [2002], the *Use of Subsea Wet-Gas Flowmeters in Allocation Measurement Systems* [API RP 85, 2002] and the *State of the Art – Multiphase Metering* [API 2566, 2004]. Full references of these publications are given in the bibliography.

This chapter intends to present the fundamentals of multiphase metering. After some general information on the basic principles and combination of elementary measurements, it describes the methods which are frequently used in commercially available meters and the usual way of reporting operational performance and accuracy. The conclusion gives overview of a number of field experiences.

3.2 FUNDAMENTALS OF MULTIPHASE METERING

3.2.1 Mixture Composition

The composition of a mixture can always be described by its volumetric fractions (or alternatively its mass fractions). The phase fractions of a mixture of gas, oil and water can be represented by a point on a ternary diagram, as shown in Figure 3.1. Two independent fractions must be known in order to describe the mixture composition. The third fraction can be derived from the requirement that the sum of phase fractions is equal to one.

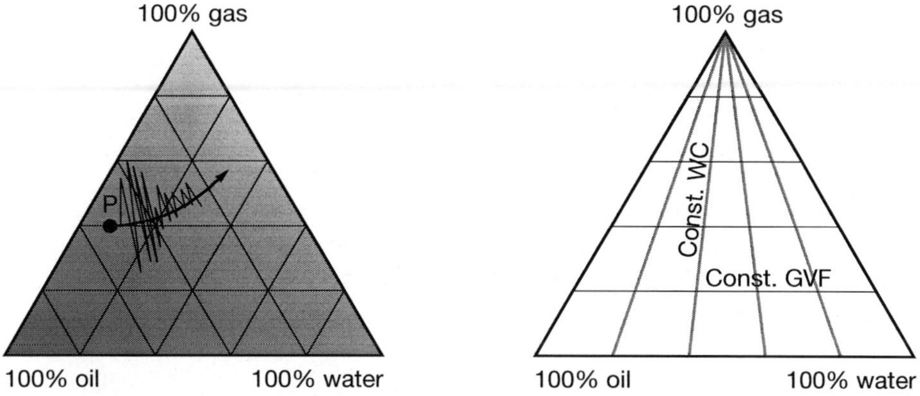

Figure 3.1

Ternary diagrams of phase fractions mapped with lines of constant phase fractions (left) or constant GVF and water cut (right)

A well stream mixture is usually described by its gas volumetric fraction (*GVF* or α_g) and its water cut w which is the ratio of water volumetric flowrate to the total liquid flowrate. The relationship between the water cut and water volumetric fraction α_w is simple:

$$\alpha_w = w\left(1 - \alpha_g\right) \tag{3.1}$$

The difference between water cut and water fraction appears clearly in Figure 3.1 where lines of constant water cut have been plotted on the right and constant water fraction on the left. Lines of constant phase fractions form three series parallel to each side of the triangle. Lines of constant water cut radiate from the 100% gas point to the opposite side.

In continuous operation, point P representing the average composition of a well stream remains localized in a relatively small area of the diagram. However, this is only true for averages taken over intermediate time periods. Over short periods (from a few minutes to days), the instantaneous flow composition may fluctuate widely, sometimes from 100% gas to 100% liquid, due to either hydrodynamic phenomena described in the first chapter or well operations (such as well restart). Over the long term (from months to years), the "average" point generally moves towards higher water fractions, and higher gas fractions.

Knowledge of this long-term evolution is important for reservoir management, but knowledge of transient measurements may be of interest to the field operators. For this reason, the ideal flowmeter will be able to follow the instantaneous fluctuations of the flow over a large area of the diagram, as accurately as possible in the main operating domain.

3.2.2 Basic Measurements

The calculation of volumetric flowrates is straightforward once the true mean axial velocities and the fractions of the cross-section filled by each phase are known. Ideally, three mean velocities (gas, oil and water) and two phase fractions should be measured.

The phase densities must also be known in order to provide the mass flowrates. In many meters, densities are not directly measured in-line, but determined periodically by sampling and testing.

Monitoring pressure and temperature conditions is also important. In crude oil production, the fluid pressure and temperature vary continuously from the reservoir to the storage tank. The phase densities and volumetric flowrates therefore vary with the measurement location, especially the gas density and volumetric flowrate. In addition, with live crude oil, the mass flowrates of the gas and oil phases are not constant. The gas dissolved in the oil phase escapes as the pressure drops, and thus the gas flowrate increases and the oil flowrate shrinks. There is a mass transfer between phases. To be correctly interpreted or compared with other measurements, flowrates must be adjusted to a well-defined reference basis (generally standard conditions) with PVT corrections for true pressure and temperature conditions, as well as gas content in liquids. Pressure and temperature measurements do not involve any particular difficulties and they are not addressed here.

Basically, a multiphase meter (MPFM) must include several independent measurements of flow properties, ideally 3 velocities and 2 phase fractions. However, direct measurement of the five required parameters is difficult or impossible, particularly the different velocities. In general, they are derived from a number of other independent measurements. The main measurement techniques used in commercially available meters are described in the following sections.

Frequently, less than five independent measurements are used. In this case, they are completed by additional relationships such as correlations of velocity slip, or by hypotheses such as assuming that the velocity slip between oil and water is negligible. In some metering approaches, additional information is extracted from the measured signal. For instance, measurement fluctuations can provide additional data concerning phase composition (see section 3.5.1).

In some cases, more than five measurements are taken, providing a degree of redundancy in the measurements which may improve the accuracy of the result. Sometimes the mean liquid density is derived from mass and volumetric flowrate measurements. Measurement redundancy must be carefully managed during data processing.

Figure 3.2 shows a diagram which describes the logical succession of steps required to derive the flowrates of each phase, for a frequently-used arrangement of measurements. Each type of MPFM has its own logical diagram, but this is not always provided by the manufacturer.

The same basic instrumentation can be used differently. For instance, the association of two capacitance measurements can give an average flow velocity which can be used to derive an estimate of the mixture density in combination with a differential pressure measurement.

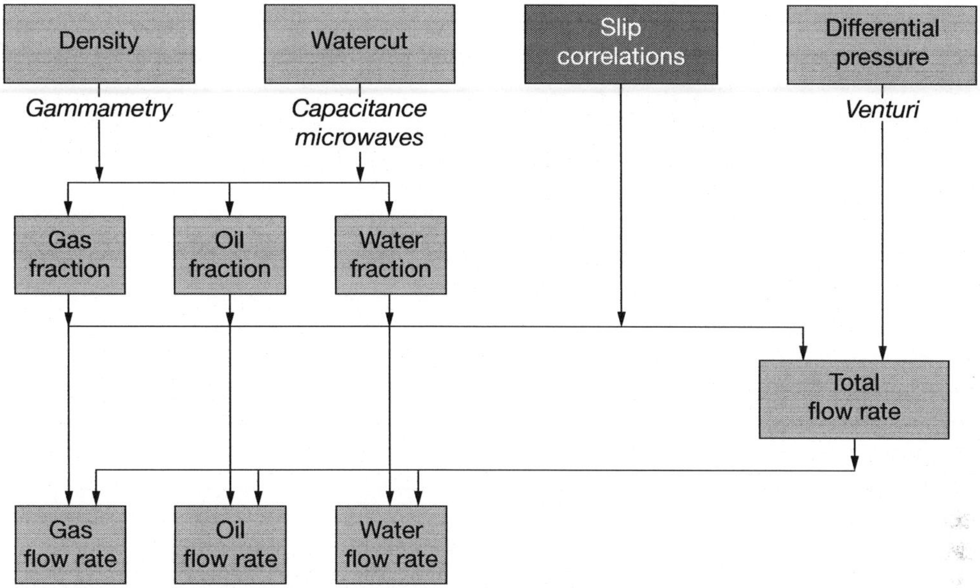

Figure 3.2

Example of combination of measurements in a multiphase meter

The combination of measurements has considerable impact on the overall accuracy of the meter. The uncertainties on flowrates resulting from a complex combination of basic measurements are difficult to evaluate theoretically and, in most cases, MPFM measurement uncertainties are determined by testing in a reference flow facility.

3.2.3 Velocity Slip Management and Flow Conditioning

Velocity slip between phases constitutes the biggest problem of multiphase metering, at least for industrial applications. Different sub-fractions of the same phase sometimes flow at different velocities. For instance, in vertical flows, small gas bubbles dispersed in liquid travel slower than large gas pockets. In lab experiments, sophisticated experimental means, like LDA (Laser Doppler Anemometry) or PIV (Particle Image Velocimetry), enable the true phase distribution of velocities to be approximated. Available means for industrial metering are limited.

Differences in true velocities are generally approximated by slip correlations. As different flow patterns may occur, this approach requires either:

- identification of the pattern and use of separate models for each case;
- or imposing a given pattern either by homogenising the flow or by using a specific position.

The first approach is difficult and at present, most of the meters commercially available use some form of flow conditioning or homogenisation. The different meters have different installation requirements, according to their design or measurement principles: horizontal, vertical upwards, vertical downwards.

In general, the velocity slip between oil and water is ignored. Tests in flow loops and field measurements show that this assumption is not always justified.

Flow conditioning often takes place upstream of an MPFM to mix and homogenise the phases and to reduce the velocity slip. Good flow homogenisation provides more regular distribution of the phase fractions in the measurement section and therefore reduces the influence of the flow pattern. Flow homogenisation also provides a more regular distribution of velocities in the measurement section, in space and over time. In such cases, the average flow may be derived more accurately from local flow measurements performed over a small fraction of the flow cross-section.

Good mixing of the multiphase flow can also reduce the velocity slip by increasing the interaction forces between phases but it does not fully cancel the slip, especially when the measuring system itself creates acceleration or deceleration of flow, as in a Venturi or orifices.

In multiphase flow, the simplest form of homogenisation system is the inverted Tee-junction pipe with the outlet oriented vertically upwards. Many multiphase meters are installed above such Tees. Other flow conditioning systems generally involve a specific internal device. They present the drawback of increasing the pressure losses, and being intrusive in the flow, and so subject to erosion or accumulation of deposits. For this reason, more recent meters generally avoid flow conditioning systems in favour of fast response instrumentation and electronics capable of processing high frequency flow fluctuations.

3.2.4 Types of Multiphase Meters

The first continuous metering on live production fluids did not use true multiphase meters but rather hybrid meters or simple systems capable of monitoring certain selected characteristics of the multiphase flow.

For example, a system with a Light Emitting Diode (LED) light source, three light sensors and a signal processing unit integrated into a package unit was used in the mid-eighties on the Bass Strait offshore field in Australia, operated by ESSO and BHP, in order to monitor the oil-in-water content in the flow at the outlet of a hydrocyclone separator [Hayes *et al.*, 1985].

This early systems certainly helped confirm the industrial benefits of multiphase metering, thus paving the way for the development of more sophisticated measurement systems, but they cannot be considered to be true MPFM.

A number of systems are still proposed with partial separation of gas and liquid. They use some kind of compact separator, generally a cyclone. Measurements are taken on the separated streams. Separation is not perfect and the liquid contains some gas and the gas some liquid, but they use sensors which can accept a limited quantity of the other phase. For instance, Coriolis flow meters can accept up to 10% of gas in liquid.

Partial separation systems are also used to extend the domain of some measurement methods to high GVF (for instance in the wet gas domain) to reduce the apparent gas content. They are also used to simplify the basic sensors. These systems are in some ways improved test separators, but they are relatively cumbersome compared to in-line models and difficult to transport.

In the following sections only MPFMs designed for continuous in-line measurement of the flowrates of the gas and liquid phases of the production stream are considered.

3.2.5 Examples of Multiphase Meters

In 2005, about seventeen different brands of MPFM were commercially available. All of them have been field tested in recent years and results have been widely published.

Table 3.1 gives examples of combinations of measurement techniques currently in use in commercially available meters. The measurement techniques are described in the following sections. The table describes the situation in 2005. Products and manufacturers are subject to change. All names given are trademarks belonging to their respective technology owners, vendors or manufacturers.

The main commercial MPFM concepts were compared in a recent paper [Falcone *et al.*, 2002] with regard to operational requirements such as measurement intrusiveness, flow conditioning and calibration needs, operating range, etc.

3.3 PHASE FRACTION MEASUREMENTS

3.3.1 Methods

This section gives an overview of the different methods used to measure the phase fractions.

The most frequently used methods are based on gamma-ray absorption and electrical measurements at various frequencies, including microwave frequencies. They both measure an average property of the multiphase mixture, density for the first, electrical conductivity or electromagnetic permittivity for the others, from which a relationship between the phase fractions is derived.

Indirect measurements of mixture densities are also derived from the combination of several types of flow measurements.

As mentioned above, two independent measurements are required to obtain the fractions of a tri-phase mixture. In commercial MPFMs, the basic methods are implemented in various combinations: gamma-ray at two energy levels, or gamma-ray and permittivity, or electrical and indirect density measurement, etc.

Table 3.1 Main commercial MPFMs

Manufacturers	Commercial Meter Names	Phase Fraction Measurement	Flow or Velocity Measurement	Comments
3-Phase Measurement AS (Framo Engineering/ Schlumberger)	PhaseWatcher Vx Phase Tester	High frequency dual gamma densitometer (Be)	Venturi	high GVF with gas extraction by phase splitter subsea version
Agar	MPFM 300 MPFM 400	indirect from mass and volume flowrates microwave	positive displacement (oval gears) 2 Venturis	high GVF with gas bypass loop
FlowSys AS FMC	TopFlow WellSense	capacitance/ conductivity and indirect from mass flowrate	cross-correlation	available sizes from 1 in. to 8 in. subsea version
Haimo	MFML 2000 L	dual gamma densitometer	cross-correlation	
Jiskoot	Mixmeter	dual gamma densitometer (^{137}Ce, ^{241}Am)	differ. pressure through a flow homogeniser	available sizes from 2 in. to 6 in.
Kvaerner Cisro	Duet	simple gamma densitometer plus a dual gamma densitometer	cross-correlation	
OVAL Corp. JOGMEC		indirect from mass and volume flowrates	twin turbine and pressure drop	
PSL	Esmer	capacitance measurements	absolute and differential pressure (Venturi)	special signal analysis available sizes from 2 in. to 6 in.
Roxar	MPFM 1900 VI	electr. impedance and simple gamma densitometer (^{137}Ce)	Venturi and cross-correlation	available sizes from 2 in. to 12 in. subsea version
TEA Sistemi Spa	Lyra	gamma densitometer impedance meter	Venturi or orifice	Vega version for wet gas

3.3.2 Gamma-Ray Densitometry

Gamma-ray absorption is probably the most common technique used in multiphase metering. Only the basic principles of gamma densitometry are described in this section. Greater detail concerning measurement principles, sources and detectors, sensor design, standards and safety can be found in a comprehensive work by Johansen & Jackson [2004].

3.3.2.1 Principles and Base of Technology

Attenuation of gamma-ray intensity depends on the density and thickness of matter and, for low-energy radiation, on the composition of matter. The gamma ray absorption technique can provide the density of a mixture flowing through a measurement system carefully calibrated at the two ends of its density measurement range.

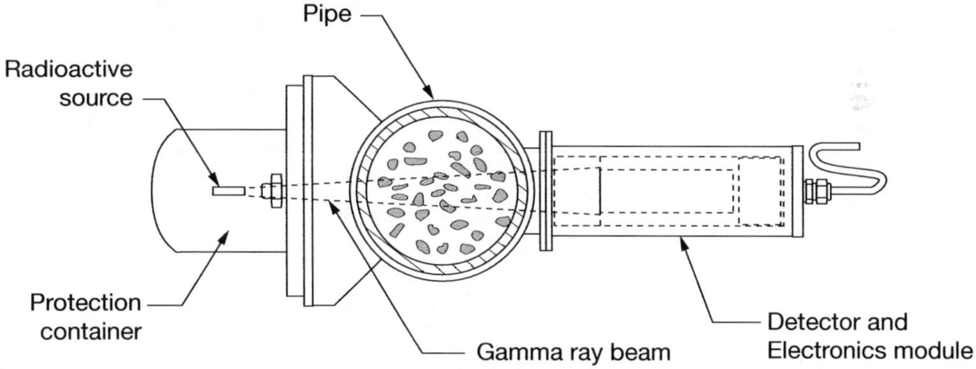

Figure 3.3

Gamma densitometer

A gamma densitometer consists of a radioactive gamma-ray source encapsulated in a protecting housing, a detector on the opposite side of the measurement section to count the particles passing through the fluid and the associated processing electronics (Figure 3.3). With a high-energy source the gamma-ray beam can pass through the steel wall and the sensor can be directly clamped onto the pipe (Figure 3.4).

The attenuation of the gamma-ray intensity is measured by counting the number of photons over a given period of time and comparing it to the number of photons emitted.

Many radioactive isotopes are available and sold for industrial applications. They are encapsulated in a sealed steel container for safety reasons. Beside cost, hazard and regulatory issues, the selection of a radioisotope is principally based on technical constraints related to its half-life, intensity and energy properties.

The half-life is the time required to observe a decay of radioisotope activity to half of its initial activity. This time must be long enough to allow practical use in an industrial application. The natural decay of source activity over time must be compensated.

Figure 3.4

Gamma densitometer installed in a test flow loop at IFP-Lyon

The intensity is related to the number of particles emitted per unit of time. A high-intensity or a long measurement time is required to limit uncertainties related to the statistical nature of particle emission. Long measurement times may involve sampling errors if the component fractions of the mixture vary rapidly. High source intensity will always reduce errors.

The energy level governs the interaction with matter and thus the penetration capability. It influences the number of particles detected and the measurement error. High-energy particles have good penetrating capability but have less interaction with matter and may be difficult to detect, requiring a longer measurement time. An optimum energy level therefore has to be found. As a radioisotope source can emit particles at several energies, with different penetration capabilities, the energy spectrum is also an important factor.

For MPFM, the radioactive source is often cesium-137 (^{137}Cs) which has a half-life of 30 years, an energy level of 662 keV and high spectral purity. A few other radioactive sources emitting high-energy gamma-rays are also used: cobalt-60 (energy level 318 keV, high spectral purity but only 5.3 years half-life), barium-133 (main energy 356 keV, 10.5 years half-life but many energy levels) and americium-241.

Many types of gamma-ray detectors exist. A good detector must have good energy resolution in order to discriminate the incident particles from other radiation and good time resolution to allow fast count rates.

In laboratories, high-purity germanium or silicon semi-conductors are often used because of their high energy resolution, but they have to be used at very low temperatures, in liquid nitrogen. Scintillation counters are preferred for industrial applications as they can be used at ambient temperature. The particle emits light detected by photo-electric cells. Sodium iodide crystals (NaI) doped by thallium ions are often used, or cesium iodide crystals (CsI) doped by sodium ions. NaI crystal, for instance, can provide an average 7% energy resolution at 662 keV (energy of ^{137}Cs) with a time resolution of 10^{-9}s.

For a homogeneous mixture, the attenuation of the gamma intensity is given by:

$$I = I_0 \exp\left[-\left(\rho_g \mu_g \alpha_g + \rho_o \mu_o \alpha_o + \rho_w \mu_w \alpha_w\right) l\right] \tag{3.2}$$

where I_0 is the intensity without the fluid but with pipe walls, and l the length of beam path in the flow. The calibration of the sensor provides I_0 and coefficients μ_i.

3.3.2.2 Single Energy and Double Energy Densitometers

Relationship (3.2) gives an equation between the phase fractions α_i. Another measurement is required to provide a second equation which, when combined with the third equation $\alpha_g + \alpha_o + \alpha_w = 1$, can be used to solve the system and find the fraction α_i, provided that the determinant of the system of equations is not zero.

In double-energy gamma-ray densitometers, the measurements are taken at two different gamma-ray energy levels instead of using an additional technique to determine the three phase fractions.

$$\begin{cases} \rho_g \mu_{g1} \alpha_g + \rho_o \mu_{o1} \alpha_o + \rho_w \mu_{w1} \alpha_w = a_1 \\ \rho_g \mu_{g2} \alpha_g + \rho_o \mu_{o2} \alpha_o + \rho_w \mu_{w2} \alpha_w = a_2 \\ \alpha_g + \alpha_o + \alpha_w = 1 \end{cases} \tag{3.3}$$

This system of equations has a single solution only if its determinant is other than zero:

$$\left(\rho_g \mu_{g1} - \rho_w \mu_{w1}\right)\left(\rho_g \mu_{g2} - \rho_o \mu_{o2}\right) - \left(\rho_g \mu_{g2} - \rho_w \mu_{w2}\right)\left(\rho_g \mu_{g1} - \rho_o \mu_{o1}\right) \neq 0 \tag{3.4}$$

If we consider a reference plane whose coordinate axes are the absorption parameters $\rho_i \mu_{i1}$ and $\rho_i \mu_{i2}$ at the two levels of energy, the above condition can be interpreted geometrically: the three points corresponding to 100% gas, 100% oil and 100% water must not be aligned in a straight line. In practice, the triangle formed by the three points must not be too flat, so that solution can be calculated with acceptable accuracy (Figure 3.5).

The above condition can be also expressed in term of the ratio of differences at the two energy levels of the absorption parameters:

$$\frac{\left(\rho_g \mu_{g1} - \rho_w \mu_{w1}\right)}{\left(\rho_g \mu_{g2} - \rho_w \mu_{w2}\right)} \neq \frac{\left(\rho_g \mu_{g1} - \rho_o \mu_{o1}\right)}{\left(\rho_g \mu_{g2} - \rho_o \mu_{o2}\right)} \tag{3.5}$$

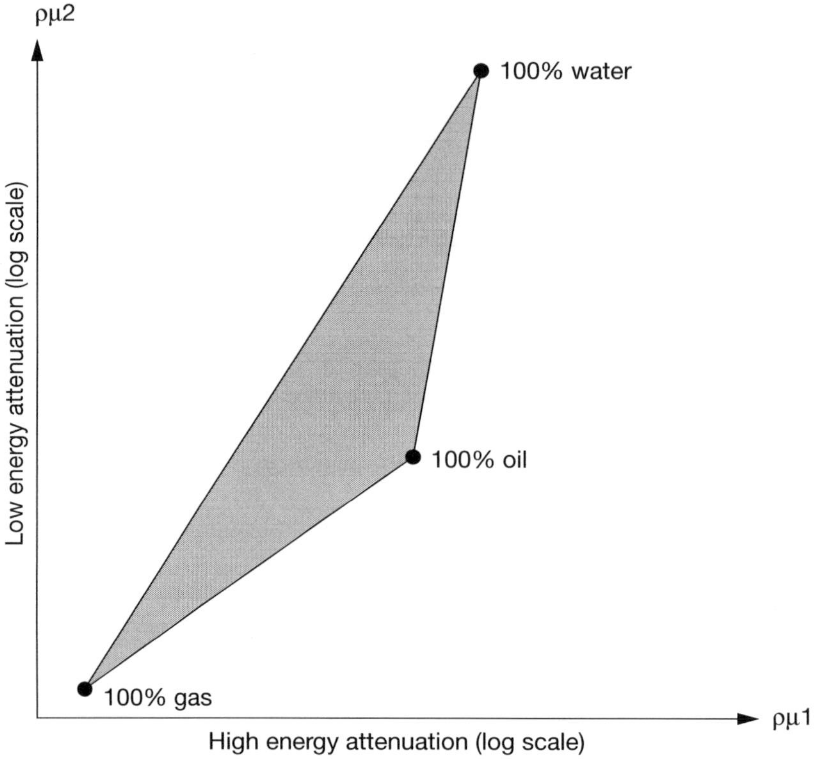

Figure 3.5

Absorption parameters at two levels of energy

The gas absorption parameter is very low and can be ignored in comparison with the others. The condition can also be written as:

$$\frac{\mu_{w1}}{\mu_{o1}} \neq \frac{\mu_{w2}}{\mu_{o2}} \tag{3.6}$$

The ratio of the absorption parameters of oil and water at the two energy levels must be sufficiently different to be able to discriminate the three components. After an experimental investigation, Nuland *et al.* [1991] reported that this condition requires a second energy level lower than 100 keV.

Two radioisotopes can be used as sources but a single one is often used at two different emission lines: for instance barium-133 (^{133}Ba) emits gamma-rays at 81 keV and 30 keV, with at 10-year half life, and americium-241 (^{241}Am) emits at energy levels of 59.5 keV and 18 keV. The detector must count the particles at two energy levels (Si detectors).

Low energy densitometers cannot be simply clamped onto the pipe, as the thickness of the steel wall absorbs gamma-rays. Gamma-ray transmission through the pipe wall must be *via* a window made of a material with a low atomic number, such as beryllium or a carbon-resin composite.

Double-energy densitometers can achieve an accuracy of approximately 2% absolute on phase fractions and 4% absolute on water cut, with a 1 second count measurement.

Water salinity influences the absorption coefficient at low energy. To overcome this difficulty, measurements at three or more energy levels have been proposed to determine the salinity concentration and make the necessary correction [Scheers, 1998]. This concept has not been marketed.

3.3.2.3 Advantages and Drawbacks of Gamma-Ray Densitometers

Gamma densitometers are independent of flow disturbances (such as emulsion or foam) and they can measure the full phase fraction range. Their accuracy is good provided that there is a sufficient contrast of density between the phases. Gamma densitometers are well-suited to gas fraction measurements.

However, they have also some disadvantages. The safe use of radioisotope elements demands a number of precautions, prescribed by the International Commission on Radiological Protection. Various national regulations also exist.

Measurement uncertainties have several origins:

- measurements are not fully independent of the flow regime, contrary to which is often claimed: this depends on the beam width.
- presence of radioactive elements in the measured mixture, such as barium sulfate scale or radium salts, can disrupt the measurements.
- external disturbances such as high temperature variations under intense sunshine.
- uncertainties due to the statistical nature of photon emission and then photon counting: measurement time must be long enough and a trade-off must be found with short sampling times to follow flow variations.
- distribution of photon energy.
- system drift.

3.3.3 Electrical Methods

3.3.3.1 General

Electrical methods are based on measurement of the average properties of the mixture: electrical conductivity or electromagnetic permittivity.

Electrical sensors are generally of simple design and are potentially less expensive than gamma densitometers. They are also easier to use (with fewer constraints than those used with radioactive components) and the acquisition time is short, allowing rapid changes in flow properties such as passage of bubbles, etc.

However, measurement depends on the distribution of phases in the measurement area and, finally, on the flow regime. In fact, gamma densitometry can be also dependent on the flow regime but this dependence is less apparent when a narrow beam is used. With electrical methods, the problem can be alleviated by an adequate sensor arrangement, especially of electrodes or antennae. A more important drawback is the influence of the phase distribution, especially whether the conductive water phase is continuous or discontinuous. This can be also managed in different ways.

3.3.3.2 Conductance

In the conductance method, the electrical current passing between two electrodes is measured. From this measurement, the global conductivity of the mixture is determined, providing a relationship between the phase fractions. Measurements are generally taken at a frequency below 100 kHz. To make the measurement possible, a continuous phase must be present and conductive. Conductance systems are therefore mainly used with water dominant mixtures, for high water cut applications. They do not work in oil-continuous mixtures. For this reason they are only used in our industry for high water cuts, generally in association with capacitance systems with a switch from capacitance to conductance when required.

3.3.3.3 Capacitance and Microwave Methods

A. Permittivity

Permittivity describes the material properties related to the transmission of an electrical field. It influences the speed (phase velocity) of the electromagnetic field. Attenuation of the electromagnetic field in the material depends on the loss factor and is often included in a complex permittivity term $\varepsilon = \varepsilon' + i\varepsilon''$.

The permittivity of a medium is often expressed relative to the free space permittivity $\varepsilon_0 \approx 8.85\,10^{-12}$ F/m. It can be measured over a wide range of freqencies. Several measurement methods exist, depending on the frequency domain. In multiphase metering, two domains are used for permittivity measurements that differentiate between capacitance (from a few kHz to a few MHz) and microwave methods (a few GHz).

Below 10 GHz, the relative permittivities of oils, and more generally the relative permittivities of non-polar liquids, are around 2. The NEL measured the permittivities of ten oils or oil mixtures over several frequencies ranging from 1 GHz to 6 GHz with temperatures between 10°C and 60°C and pressures up to 400 bar, under a project funded by the U.K. National Measurement System Directorate [NEL, 1999]. The measurements were fitted to an analytical formula dependent on temperature and pressure. The relative permittivities were found to be only slightly influenced by the frequency over the measurement range and were linked by a linear function to the liquid density in the same way as for many non-polar liquids. Previously, Marelli *et al.* [1992] related the relative permittivity of crude oils to their API degree, *i.e.* to their density.

The relative permittivities of gas mixtures are near 1. They are influenced by the molar density [St-Arnaud *et al.*, 1997]. The permittivity of water and brines versus frequency are given in several textbooks, for instance that published by Ulaby *et al.* (1986).

Figure 3.6 shows the theoretical average relative permittivity of an oil-water mixture at high frequency, for droplets of water in oil and droplets of oil in water respectively. However, as the inversion point of the emulsion usually occurs over a relatively narrow range of water cuts, with a clear transition, the large permittivity difference enables the two types of mixture to be differentiated. For instance, in Figure 3.6 a relative permittivity measurement above 20 characterizes droplets of oil in water and below 20 characterizes droplets of water. There is a jump in relative permittivity when the inversion point of emulsion is crossed.

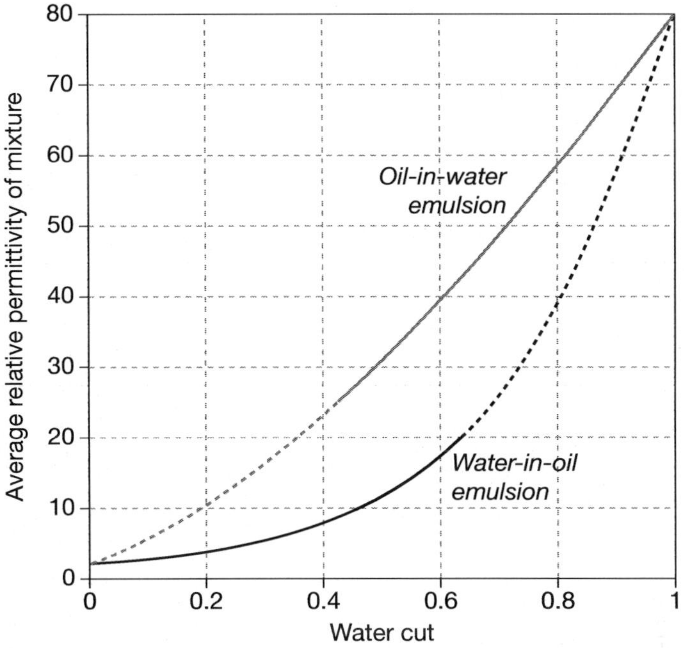

Figure 3.6

Influence of the continuous phase on the average permittivity of an oil-water emulsion.

B. Capacitance

The capacitance method consists in measuring the electrical capacity (imaginary part of the impedance signal) between two electrodes, with an alternative current of a frequency chosen between few kHz and few MHz.

The measured capacity depends on the average fluid permittivity between the electrodes. It depends on the electrode configuration but also on the distribution of phases. The measurements are not identical, for the same phase fractions, when the flow is stratified, annular, dispersed with an oil-continuous phase or dispersed with a water-continuous phase.

To be able to interpret the measurements correctly, the flow pattern must be either known or imposed with a mixing device upstream of the metering section or a special piping configuration (Tee-junction for instance). Suitable electrode design can also reduce the influence of the flow pattern. For instance, vertical strip electrodes are less sensitive to stratified flow than horizontal electrodes.

Several electrode arrangements are used. The main arrangements are two ring electrodes and two or more strips lined up with the flow direction, but helical electrodes have also been tested [Tollefsen, 1998]. Frequently, dummy electrodes are added to smooth the electromagnetic field and reduce the influence of flow pattern. In oil flowmeters the electrodes generally have the shape of a half cylinder covered by a ceramic tube liner, or are directly in contact with the fluids.

C. Microwaves

As with the capacitance method, microwave measurements are based on the difference of electromagnetic permittivity between the different phases, but the measurements are taken at much higher frequencies, in the 0.8 to 10 GHz band. The upper limit (10 GHz) is a practical one, to avoid excessive absorption by water and also to avoid the high cost of electronic components.

Over the above frequency domain, the relative electromagnetic permittivities of the fluids range from $\varepsilon = 1$ for gas to around $\varepsilon = 2$ for oil and up to $\varepsilon = 80$ for water. There is a significant contrast between the relative permittivity of water on the one hand and oil and gas on the other, which allows easy detection of water. For this reason, microwave methods are primarily applied to water cut measurements. However, microwave measurements have been also applied to three-phase mixtures [Marelli *et al.*, 1992].

Two main approaches are used in MPFM:

- cavity resonance,
- measurements of attenuation and phase shift during microwave propagation through the studied media.

Cavity resonance is laboratory method commonly used measure the permittivity of low-loss materials at microwave frequencies (see for instance reference NEL [1999]). The sample material is placed in a closed cell and exposed to a microwave field. A peak in the wave amplitude can be observed at a given frequency, depending on the material permittivity and size of the cell. The shift in frequency from a reference frequency, for instance when the cell is empty (full of air) characterises the material permittivity. The method has been adapted to a pipe section, using an active microwave circuit, self-oscillating at the resonance cavity to avoid scanning over a range of frequencies that would take too long.

In the second approach, the amplitude and phase shift of the electromagnetic field is measured after transmission though the fluids. The received signal is compared with the transmitted signal through a reference line in order to determine the attenuation and the phase shift, to eliminate possible fluctuations in the electronics. After preliminary processing which eliminates the effects of wave propagation outside the fluid mixture, the phase shift is directly the consequence of the average relative permittivity ε_m of the mixture (for 100% gas $\varepsilon_m = 1$, if gas is used as reference) using the following formula:

$$\Delta\phi = \frac{2\pi f d}{c_0}(Re(\sqrt{\varepsilon_m}) - 1) \tag{3.7}$$

where $\Delta\phi$ is the phase shift, f the microwave frequency, d the distance between the transmitter and receiver sensors, c_0 the velocity of light in air. This relationship is approximate because the effects of wave reflection on wall and interfaces are ignored. However, errors cans be minimised if the sensor is well-designed. A second relationship can be established between the signal amplitude and the average $Im(\varepsilon_m)$. The effect of wave reflections is more important on amplitude than on phase shift, and this is influenced by fluid permittivity. This influence can be described by a transfer function which can be fitted by calibration to a polynomial relationship.

At first glance, it might appear that only one parameter is required to calculate the average mixture permittivity, such as the phase shift which leads to a simple and more accurate relationship. Unfortunately, phase shift measurements are made *modulo* 360 dg, while the range of possible phase shifts given by Equation (3.7) generally exceeds 360 dg, except if the fluid thickness d and the microwave frequency f are small. This implies that two, three and sometimes more solutions are possible. For instance, if the measured signal gives a 70 dg shift, permittivities corresponding to 430 dg or 790 dg could also be a solution. Simultaneously taking account of the signal attenuation enables the correct permittivity to be selected, even though this attenuation measurement is less accurate.

The next processing step consists in deriving the phase fractions from the average mixture permittivity. They can be related by a mixture law. For instance, the generalised Bruggemann-Hanaï law is well-suited when one component is polar, like water:

$$\frac{\varepsilon_m - \varepsilon_i}{\varepsilon_h - \varepsilon_i}\left(\frac{\varepsilon_h}{\varepsilon_m}\right)^L = 1 - p_i \tag{3.8}$$

where ε_i is the permittivity of inclusions, ε_h the permittivity of the continuous phase, p_i the volumetric fraction of inclusions and L a form factor. The continuous phase could be either oil or water.

The Bruggemann-Hanaï mixture law is unsuitable for gas-oil mixtures, but other laws, such as the CRIM (Complex Refractive Index Method) mixture laws can be used. The CRIM mixture law is expressed as follows in its generalised form for n components:

$$(\varepsilon_m)^c = \sum_{i=1,\ldots,n} p_i(\varepsilon_i)^c \tag{3.9}$$

where p_i is the volumetric fraction, ε_i the permittivity of component i and c form parameters.

To apply the microwave method to mixtures composed of three phases, Bentolila and Constant [2005] considered that the mixture was a combination of two two-phase mixtures. The first one is a constant *GVF* model that can be interpreted as a mixture of water and a hypothetical fluid of permittivity ε_{GVF} which is itself given by a combination of the two phase oil-gas mixture law. This constant *GVF* model is used to estimate the water fraction p_w. Similarly, a constant water cut model can be interpreted as a mixture of gas and liquid which is itself a combination of oil and water. This constant *WC* model is used to estimate the gas fraction p_g (or GVF). The three phase fractions can be obtained by scanning all the two-phase model intersections on the map area adjusted with the right conditions of temperature and salinity.

3.3.4 Indirect Density Measurements

The density of the multiphase mixture is sometimes derived indirectly from several flow measurements, such as the combination of volumetric flowrate and differential pressure. Coriolis flowmeters, well-known in single-phase flow measurements, provide the average fluid density simultaneously with its mass flowrate. They can be used also in multiphase flow, but over a limited range of gas fractions.

The following paragraphs are limited to application of flow measurements for calculating the mixture density. Details on the corresponding methods are given in section 3.4 of this chapter.

3.3.4.1 Coriolis Meters

In Coriolis meters, the fluid flows in a U-shaped pipe forced to vibrate by a driver. The phase shift of vibrations, relative to vibrations in no-flow condition, is proportional to the mass flowrate. The natural vibration frequency depends on the fluid density. This property enables the fluid density to be determined when the pipe vibrates at its natural frequency.

The response of multiphase fluids to accelerations is complex due to the various interactions between the phases (drag, inertia and Basset effects). The above properties, established in single-phase fluids, are no longer true. However, Coriolis meters can still be used with acceptable accuracy in bubble flows at low gas fractions (lower than 5%). Applying a correction to the apparent gas fraction enables the domain to be extended up to gas fractions of 15% with 5% accuracy [Adejuyigbe *et al.,* 2004]. This accuracy is not sufficient for differentiating between oil and water.

3.3.4.2 Combination of Differential Pressure and Volumetric Flowrate

If the volumetric flowrate is measured simultaneously with the pressure drop in a differential pressure device, the average homogeneous density ρ_h can be derived from the orifice relationship given below, re-arranged as [Lin, 1980]:

$$\rho_h = k^2 \frac{2\,\Delta P_{TP}}{Q_v^2} \tag{3.10}$$

The relative error on the homogeneous density is given by: $e_\rho = e_{\Delta P} + 2\,e_Q + e_k$. This equation shows that a high level of accuracy must be obtained on the volumetric flowrate measurement.

In multiphase flow, however, the main problem is related to the influence of the true phase fractions on factor k which is detailed below. The velocity slip is directly introduced into the relationships derived from the differential pressure measurements and this additional parameter is eliminated. This enables the gas fraction to be determined.

3.3.4.3 Combination of Two Differential Pressure Measurements

Using this approach, two differential pressure devices are mounted in series to perform two measurements of the same mass flowrate. The response of the two sensors must differ sufficiently according to the gas fraction to be able to discriminate the influence of this parameter. The method is detailed by Lin [1980] who indicates that the error on flowrates can reach 30%.

To overcome the difficulties presented by this and the previous approach, one of the leading MPFM manufacturers combines them, i.e. two different Venturi meters are mounted in series with a volumetric meter. The velocity slip is introduced as a parameter into the relationships provided by the two differential pressure measurements in addition to the gas fraction. The two differential pressure measurements allow elimination of the slip and solving of the gas fraction with the volumetric measurement [Agar, 2002].

3.4 FLOW MEASUREMENTS

3.4.1 Methods

As for phase fractions, various methods can be used to measure flowrates in multiphase flow. The most popular method is to measure pressure differences through a restriction, such as a Venturi or a calibrated orifice. Volumetric measurement devices are also used, such as turbines or screws. Application of ultrasonic methods have been investigated in multiphase flows but active approaches have not yet succeeded in producing a commercial system. However, passive acoustic sensors are used successfully in cross-correlation methods and in advanced "pattern recognition" methods.

3.4.2 Differential Pressure Measurements

In single phase flow, the mass flowrate Q_M through a Venturi tube or any contraction accelerating the fluid (orifice plate, etc.) is related to the differential pressure ΔP by the following expression, derived from Bernoulli's law:

$$Q_M = A_{\min} C Y \sqrt{\frac{2\rho\,\Delta P}{1-\beta^4}} \tag{3.11}$$

where: $A_{\min}$ Venturi throat area ($A_{\min} = \frac{\pi d_{\min}^2}{4}$ with $d_{\min}$: diameter at throat),

β diameter ratio ($\beta = \frac{d_{\min}}{d}$ with d: upstream pipe inside diameter),

C discharge coefficient, a function of the Reynolds' number,

Y expansion coefficient, accounting for gas compressibility effects,

ρ fluid density,

ΔP measured differential pressure, the hydrostatic contribution $\rho g \Delta z$ must be removed if the Venturi is not horizontal.

Equation (3.11) shows that Venturi and orifice measurements require knowledge of the fluid density in order to derive either the mass or volumetric flowrates. The measured differential pressure is proportional to the product of these two flowrates and they are to a certain extent "in between" mass and volumetric flow meters.

In two-phase flow, if we assume that the fluid is a homogeneous mixture with no velocity slip between the different phases $(V_g = V_o = V_w)$, the total mass flowrate is still given by equation (3.11) but the fluid density ρ must be replaced by the "homogeneous density" of the mixture:

$$\rho_h = \alpha_g \rho_g + (1 - \alpha_g)(1 - w)\rho_o + (1 - \alpha_g) w \rho_w \tag{3.12}$$

where α_g is the volumetric gas fraction and w the water cut.

However, flowrates calculated in this way show significant discrepancies with measurements. The convective acceleration in the convergent section locally disturbs the flow. The gas is accelerated more than the liquid phases and a velocity slip is therefore created in the throat section, even though the mixture is perfectly homogeneous at the Venturi inlet. The phase distribution in the cross-section is also modified, with a greater concentration of gas in the centre.

The measured pressure drop ΔP_{TP} is generally lower in two-phase flow than in single-phase flow of the same mass flowrate and the same homogeneous density. It follows that, for a given pressure drop measurement, the total mass flowrate is overestimated by equation (3.11). To deal with these effects, a two-phase corrective multiplier K_{TP} is generally introduced into the equation:

$$Q_M = A_{\min} C Y K_{TP} \sqrt{\frac{2\rho_h \Delta P_{TP}}{1 - \beta^4}} \tag{3.13}$$

The gas mass flowrate Q_{Mg} is then derived, provided that the the gas mass fraction χ_g is known. This is also called the "quality" of the mixture:

$$Q_{Mg} = \chi_g Q_M = \frac{\alpha_g \rho_g}{\alpha_g \rho_g + (1 - \alpha_g)\rho_l} Q_M \tag{3.14}$$

The corrective multiplier K_{TP} can be determined experimentally by calibration in a multiphase flow test loop. It depends both on the Venturi geometry and on flow parameters such as GVF, water cut, flowrates (or superficial velocities) and phase density ratios. The viscosity of the liquid has also an influence on the measurement. It is sometime combined with the discharge coefficient C.

The liquid or gas density can be used as in equation (3.11) instead of using the unknown "homogeneous" mixture density. For wet gas or steam measurements which were the principal applications in the past, the gas density is preferred as a reference. The gas mass flowrate can be written:

$$Q_{Mg} = A_{\min} C Y K_g \sqrt{\frac{2\rho_g \Delta P_{TP}}{1-\beta^4}} \tag{3.15}$$

with the following relationship between the coefficients K_g and K_{TP} :

$$K_g = \chi_g K_{TP} \sqrt{\frac{\rho_h}{\rho_g}} \tag{3.16}$$

If the single-phase mass flowrate of gas producing the pressure drop ΔP_g is taken as a reference:

$$Q_{Mg} = \phi_g Q_{g\,\text{ref}} = \phi_g A_{\min} C Y \sqrt{\frac{2\rho_g \Delta P_g}{1-\beta^4}} \tag{3.17}$$

and

$$\frac{Q_{Mg}}{Q_{g\,\text{ref}}} = \Phi_g = \frac{\varepsilon_g \rho_g}{\varepsilon_g \rho_g + (1-\varepsilon_g)\rho_l} K_g \sqrt{\frac{\Delta P_{TP}}{\Delta P_g}} \tag{3.18}$$

Several corrective two-phase flow multipliers may be found in the technical literature but they were generally derived from measurements through orifice plates using steam and water and for volumetric gas fractions R_g larger than 97% or 98%, i.e. in the wet gas domain. An overview of published correlations was given by Lin [1980]. Few correlations are available for Venturis at medium and low gas fractions. Correlations used in commercial meters are generally confidential proprietary data.

The available correlations include those given by Murdock [1962], Chisholm [1977], de Leeuw [1997], Hall *et al.* [2000] and Steven [2001]. The last three are established with Venturis and not with orifices.

Murdock used a simple corrective multiplier which depended only on the Lockhart-Martinelli parameter X[1]:

$$K_g = \frac{1}{1+1.26X} \tag{3.19}$$

This multiplier must be applied using formula (3.15), with the gas density under the square root and gives an estimate of the gas mass flowrate. In general, care must be taken to ensure that the appropriate reference density and flowrate are used. A factor of 1.26 was established by fitting to experimental data over the range of variation of the Lockhart-Martinelli parameter between 0 and 0.30.

1. The Lockhart-Martinelli parameter is the square root of the ratio of liquid and gas single phase pressure gradients which would be obtained if each fluid was flowing alone at the same superficial velocity as under two-phase flow conditions $X = \sqrt{\frac{dP_{Ls}/dx}{dP_{Gs}/dx}}$.

Chisholm assumed a constant slip ratio between the gas and liquid velocities. This assumption led to a corrective multiplier having the form: ,

$$\Phi_{TP} = \frac{1}{\sqrt{1 + CX + X^2}} \tag{3.20}$$

where the parameter C is a function of the density ratio ρ_g / ρ_l and has two different expressions depending on whether X is lower than or greater than 1.

De Leeuw used the same type of corrective multiplier as Chisholm, but his constant C is a function of a modified gas Froude's number $\mathbf{Fr}_g = \frac{V_{Sg}}{\sqrt{gD}} \sqrt{\frac{\rho_g}{\rho_l - \rho_g}}$ (where V_{Sg} is the gas superficial velocity and D the diameter). This modification was based on tests using a Venturi and hydrocarbon fluids.

Steven [2001] carried out a series of tests with a Venturi and compared the various corrections indicated above. He found that de Leeuw's correction was the more accurate for this type of flow restriction. He also proposed a modified correlation, which was more accurate but somewhat too lengthy to be presented here. Details can be found in the given reference.

Corrections for Venturis of several diameter ratios β have been published by Hall *et al.* [2000] over the full range of GVF. Their corrections are based on the true gas fractions and the corresponding average densities.

The choice of the diameter ratio β of the Venturi is particularly important in two-phase flow. A low β increases the pressure drop and the sensitivity of the measurement. However, accelerations are also increased, with a negative impact on the two-phase corrective coefficient K_{TP}. A trade-off must be found by the designer.

3.4.3 Volumetric Meters

Volumetric meters directly give the total volumetric flowrate of fluids. There is a difference between positive-displacement meters and turbine meters. Both types are well-known technologies in single-phase flow and have been applied to multiphase flow metering.

3.4.3.1 Positive-Displacement Meters

Positive-displacement meters are based on the successive filling and emptying of chambers of known volumes. In many designs the chambers are continuously created by the rotation of a mechanical device, the rotation speed of which gives the total volumetric flowrate.

Several types of meters such as these are used with both liquid and gas single-phase flows: helical gears, oval wheels, sliding vanes, etc. Helical gear meters and oval wheel meters have been used in multiphase meters, particularly the second type.

Positive-displacement meters can offer high accuracy but they can be sensitive to solid particles and wear. They are also relatively bulky for large flowrates. In single-phase flow, they are better suited to fluids of medium to high viscosity while turbine meters are preferably used with low viscosity fluids.

3.4.3.2 Turbines

In single-phase flow, turbine flow meters have considerable accuracy and are extensively used when accurate measurements are required. The rotation speed of the turbine is proportional to the volumetric flowrate. Correction parameters may account for liquid viscosity effects.

They are also used for multiphase metering but are intrusive in the flow and may be subject to wear and erosion by sand particles. For this reason, they are less widely used than static devices.

In the same way as for differential pressure devices, a correction factor which is a function of GVF and water cut must be introduced. The rotation speed decreases almost linearly when the GVF increases and to a lesser extent with increasing water cut. The accuracy relies fully on this correction factor.

Dual turbines can provide data on the total mass flowrate if they are coupled by a torsion spring. The turbine blades must have different angles of attack to rotate freely at two different rotation speeds. The differential angle between the two turbines is proportional to the total mass flowrate [Li *et al.*, 2004].

3.4.4 Cross-Correlation

In the cross-correlation technique, two identical sensors are installed at some distance along the pipe. The sensors can be of any type as long as they do not disturb the flow. Velocity measurements have been performed by cross-correlation with impedance sensors, capacitance sensors, microwave sensors, etc.

The two measured signals are processed by cross-correlation to identify the prevailing time lag for propagation of flow disturbances (turbulent pressure fluctuations in single-phase flow, gas bubbles or pockets, liquid slugs, etc. in two-phase flow).

Cross-correlation is well-adapted to slug flow or flow with large gas bubbles. It gives the velocity of the slug or gas bubble. The slug velocity is not exactly the same as the liquid superficial velocity. It is about 20% higher, but the ratio depends on the superficial velocity and increases at low superficial velocities.

3.4.5 Ultrasonic Measurements

Since the 80's, the use of ultrasonic sensors has grown considerably in single-phase flow metering owing to their performance, compactness and accuracy. In spite of this development, ultrasonic techniques are seldom applied to multiphase flows. Several technical issues hinder their application to MPFM:

- background noise,
- large variations in the acoustic impedance of the surrounding media,

- signal attenuation by diffusion in heterogeneous media (bubbles, droplets),
- variation in sound velocity with flow regime and GVF.

Ultrasonic measurements can be passive or active. In passive systems, pressure fluctuations or noise resulting from the turbulent structure of the flow are measured and analysed either by a cross-correlation method, as explained in the previous section, or by other advanced methods (see section 3.3.5.1.). The mixture velocity must be high enough to be in the turbulent domain. This is why the Reynolds number must be above the critical Reynolds number.

In active systems, a transmitter generates an acoustic signal which is modified during its propagation in the multiphase mixture. The received signal is analysed to determine the velocity of the flow. Two main principles with a number of options are used in ultrasonic flowrate measurements:

- measurement of apparent sound velocity in a moving medium (transit time or time of flight),
- analysis of signals back-scattered from particles flowing within the fluids.

Active ultrasonic measurements offer good potential for direct measurement of phase velocities, but at present, no single type of sensor or method cannot cover the full operating domain.

3.4.5.1 Acoustic Transducers

Acoustic transducers are sensors which transform acoustic waves into electrical signals and, conversely, electrical signals into acoustic waves. Generally, they are based on three physical principles: piezo-electricity, electromagnetism and electrostatics. The selection is made according to the particular case.

A good transducer must be able to transfer acoustic energy from or towards the medium efficiently, to offer good sensitivity, wide bandwidth, good impulse response and good directivity. This is achieved when designing the sensor by optimising selection of the following parameters: working frequency, type of transducer, shape and sizes, materials, electrical impedance. The type of measurement influences the design (transit time, Doppler, etc.). The sensor must also be able withstand its environment, that is the fluid pressure and temperature. In single-phase flow, the influence of the various parameters is well-known and they can be chosen to guarantee efficient transfer of acoustic energy.

Existing standard transducers are designed for single-phase flow, at relatively low pressure and low temperature. They are not suitable for multiphase hydrocarbon media which have a highly-variable acoustic impedance. For multiphase mixtures, the design of specific transducers is generally more complex, as it must balance several conflicting constraints such as acoustic properties, environmental aspects and safety issues. To ensure good performance, these sensors generally include a backing part, a piezo-electric ceramic and an impedance-tuning plate in front of the fluid. These elements are usually enclosed in a housing. The dimensions and acoustic impedance of the backing and the plate also have considerable influence on the sensitivity and bandwidth of the sensor. For these components, selection of matcrials is of paramount importance. In general, they are made of a composite

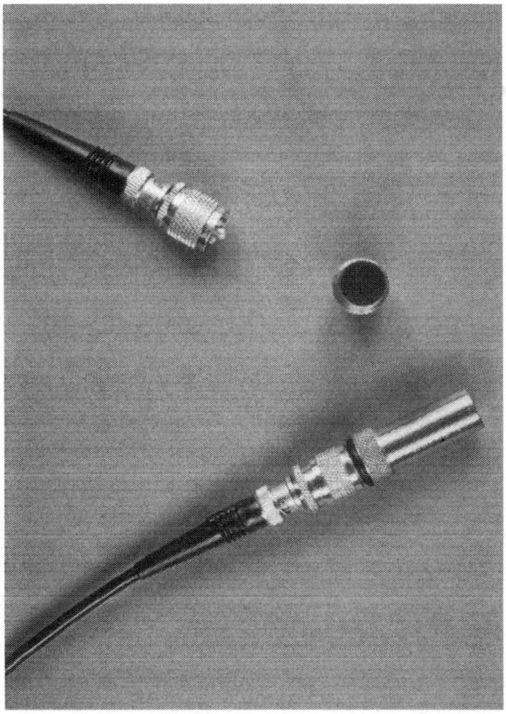

Figure 3.7

Typical acoustic transducers

material suitable for the pressure and temperature conditions and the characteristics of the fluids. Selection of the piezo-electric ceramic is mainly influenced by its electro-acoustic performance and by its temperature resistance. As the sensors are in contact with flammable fluids and used in hazardous areas, they must comply with appropriate regulations and satisfactorily undergo safety testing.

3.4.5.2 Transit Time Measurements

With the transit time method, two sensors are installed at some distance along the flow, in opposite positions. They are alternately transmitter and receiver. The flow velocity is derived from the apparent sound velocity measured in the downstream and upstream directions (pulse transit time divided by sensor distance). The half difference provides the flow velocity V and the half sum the sound velocity c. It can be easily established that the relative error on flow velocity $e_V = \Delta V/V$ depends on the measurement error e_c on the apparent sound velocity divided by the Mach number of flow (i.e. V/c):

$$e_V \cong e_c\, c/V \,. \tag{3.21}$$

To be accurate, the method requires a Mach number large enough. Usually, liquid and gas flows in pipes have very low Mach numbers. Fortunately, in low-pressure conditions, the sound velocity falls below 150 m/s in gas-liquid homogeneous mixtures over a wide range of intermediate GVF, but it is greater either for small or very large GVF and for high pressures.

A second important limitation of acoustic measurements comes from signal attenuation by diffusion. Attenuation depends on liquid viscosity and bubble size. It also depends on the signal frequency. Lower frequencies give weaker attenuation but they require bulkier transducers and imply less resolution.

3.4.5.3 Acoustic Signal Backscatter

The acoustic wave generated by an acoustic transducer is modified by reflection, diffraction, diffusion and attenuation in a moving fluid. Acoustic diffusion occurs when the fluid is charged with diffusing particles. The diffused signal is received by the transmitting or another transducer. The velocity of particles can be derived by two methods: time lag measurements and Doppler effect. The derived velocity can be considered equal to the fluid velocity if the particles are small enough.

The time lag method uses pulsed acoustic waves. The measurement of the time lag difference between reception of two backscattered pulses and transmission enables the velocity of moving particles to be determined.

The second method, based on the Doppler effect, is well-known using both acoustic and electromagnetic waves. When an incident wave hits a moving particle, the reflected wave is shifted in frequency. The frequency shift is proportional to the particle velocity. Several options are available: continuous Doppler, vertical or transverse pulsed Doppler, etc.

The time lag and Doppler effect methods both depend on sound velocity, which is strongly influenced by the phase fractions in multiphase flow and generally not known. This is a major source of measurement uncertainty. The sound velocity must therefore also be measured directly to ensure sufficient accuracy. Depending on the method, the relative error on flow velocity is given by:

$$e_V = e_c + 2\, e_{\Delta t} \text{ or } e_V = e_c + 2\, e_{\Delta f} \qquad (3.22)$$

where $e_{\Delta t}$ is the relative error on time lag measurements and $e_{\Delta f}$ that on frequency shift estimations. As the diffusing particles do not flow with the same apparent velocity, the measured signals must be processed to determine the flow velocity. For instance, the power spectral density of the signal must be calculated to estimate the frequency shift. The autoregressive filters technique provides a relatively simple means of estimating the frequency shift. Several other disturbance factors spread out the spectral density and have an influence on the relative accuracy $e_{\Delta f}$, for example the aperture angle of the acoustic beam or the surrounding turbulence noise.

Accuracy is generally better than with transit time measurements but there are still limitations duc to signal attenuation.

3.5 OVERVIEW OF ADVANCED METHODS

3.5.1 Analysis of High Frequency Flow Signal

Measurements in natural multiphase flows always shows high-frequency random signal fluctuations produced by turbulence and instantaneous variations of phase fractions. Some approaches make use of the characteristics of these fluctuations (intensities, spectral frequency components, etc.) to derive individual flowrates. These are often referred to as "pattern recognition" approaches. The objective is to minimise the number of sensors, their complexity and finally the cost of the meter.

Depending on the type of sensors, phase fractions and velocities can be measured by pressure fluctuations, by variations in electrical properties or even by variations in density using gamma rays. A combination of several techniques is used in some systems. Typically, the scanned frequency band is between 25 and 500 Hz. The sensors are identical or similar to the sensors described above, but the measured signal is extensively analysed to extract more information. Van der Speck and Thomas [1998] have shown that a neural network approach can correctly identify the flow regime by using band spectra of flow-generated sounds.

Figure 3.8

Esmer Flowmeter by PSL under testing at IFP-Lyon

Already marketed by PSL, the "Esmer" meter is based on analysis of high-frequency signals emitted by standard sensors to determine the flowrates of each phase [Toral *et al.*, 1998]. The measurements only include differential and absolute pressure measurements though an orifice and impedance measurement. The sensors are assembled in a straight-line spool piece (Figure 3.8). Signal analysis is not based on a physical model, or an empirical model, but on a neural network approach. The calibration of the system (learning phase) is based on a combination of laboratory tests (or tests in an industrial multiphase loop) completed by field tests on-site to adjust the response to the true operating conditions.

A Canadian flow meter, the "V-H meter", under development at the University of Alberta, is based simply on analysis of pressure fluctuations measured at the ends of two sections of a metering loop, a vertical one and a horizontal one [Toma *et al.*, 2000]. Pressure fluctuations result from fluid accelerations produced by local void fractions. The two measured signals are filtered to remove unwanted noise. Their power spectral density functions are then calculated to derive the phase velocities. This approach leads to a relatively simple and cheap device but, for the time being, one that requires a relatively long sampling time (around 2 mn) to obtain power spectral density functions that are stable enough. It is also perhaps a little less accurate than more conventional meters, nearer to 10% accuracy for liquid and gas flowrates than 5%.

A multiphase metering device proposed in 2004 by a small Norwegian company, Abbon, uses a single passive acoustic sensor, from which liquid and gas flowrates are derived by proprietary signal processing. Acoustic noise must be generated upstream of the sensor by a flow conditioner or a choke. This approach could lead to a particularly low-cost device but no sensor performance results have as yet been released by the vendor.

3.5.2 Microwave Doppler Velocity Measurements

Doppler velocimetry is now widely used in many areas. Police radars or laser guns for measuring the speed of cars are now familiar to the general public. Rain radars for weather forecasting and measurement of ball speeds in tennis, golf and baseball, are also well-known applications of Doppler electromagnetic velocimetry.

In spite of a small ratio of the flow velocity relative to the electromagnetic wave celerity, the velocity can be derived with good accuracy as the time lag can be precisely measured.

A patent has been applied for concerning an application to well stream flow metering [Wu *et al.*, 2000]

The main issues for application of the Doppler method to petroleum media are:

- signal attenuation in conductive media (water, especially with possible presence of scale);
- the need for wave reflectors in the flow (such as interfaces or particles) with a size compatible with the wave length (not too small to avoid diffusion phenomena);
- variable wave celerity depends on fluid composition. In fact, the wave velocity can be measured directly using the phase lag of the transmitted and received signal.

3.6 PERFORMANCE DESCRIPTION AND CALIBRATION

3.6.1 Operating Domain

A measuring instrument is characterised by an operating domain or range over which it can provide a measured value with a specified accuracy. Multiphase meters give several measurements which are not independent of each other. For instance, liquid and gas flowrate accuracy can be influenced by the gas fraction. The notion equivalent to the range is a multi-dimensional envelope.

The flowrate, is usually presented on a two-dimensional map, with axes graduated respectively in gas (actual conditions) and liquid flowrates, in logarithmic scales, or alternatively in superficial velocities of gas and liquid (Figure 3.9). In this type of diagram, there is no distinction between oil and water flowrates. Lines of constant gas fraction are diagonals rising from left to right. The various operating domain boundaries can be related to various limits which impact the final measurements. They depend on the basic types of measurements or the performance of a specific sensor, for instance minimum or maximum flow velocities, GVF measurement, etc.

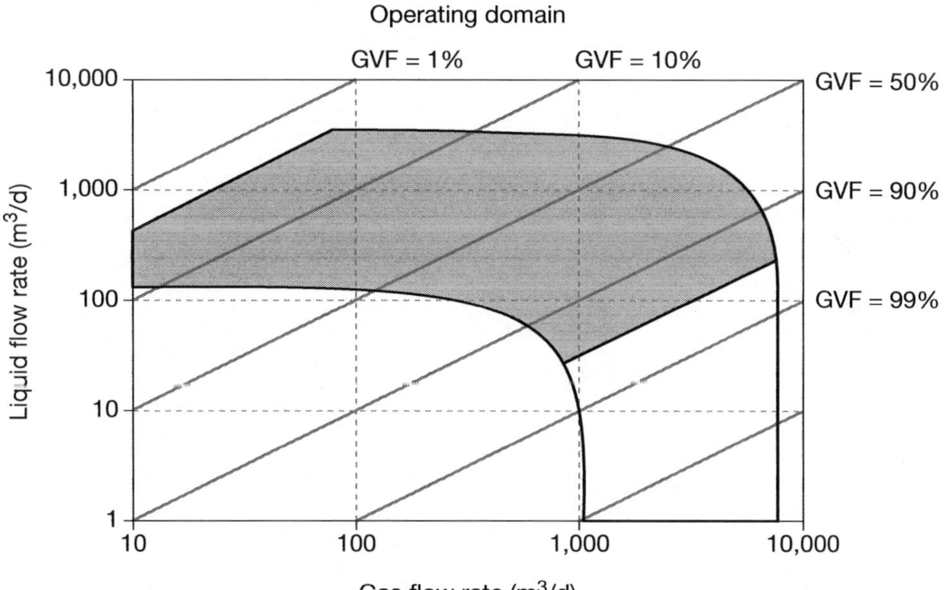

Figure 3.9

Typical chart of an MPFM operating domain

The volumetric fractions are significant parameters for an MPFM operating domain. This domain is also often presented in another or additional form, in diagrams of gas fractions versus water cuts as shown on Figure 3.10 (on left). The use of a ternary diagram gives a similar alternative representation (Figure 3.10, on right).

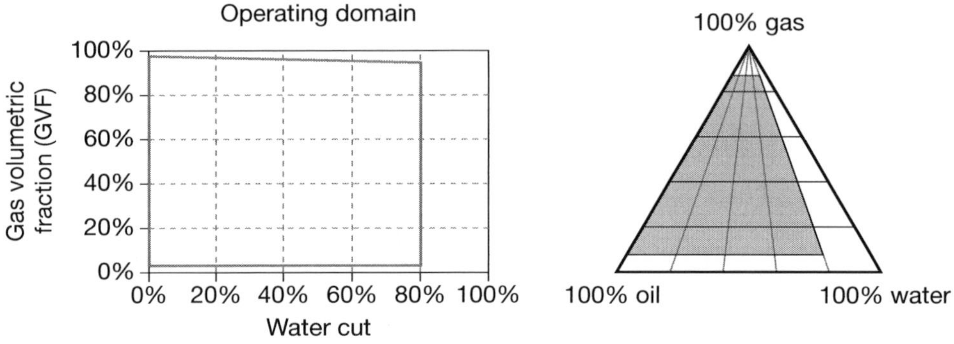

Figure 3.10

Typical GVF-water cut charts

3.6.2 Performance Description

The Performance of an MPFM is described in term of accuracy, repeatability and sensitivity or tolerance.

3.6.2.1 Accuracy

Accuracy characterises the ability of a measuring instrument to provide a measured value that is close to the true value. The difference between the measured value and the true value is the measurement error. It may be expressed either directly by the difference (absolute error) or by the ratio of this difference to the measured value (relative error).

For multiphase meters, the above definition must be applied to each phase. It is also sometime applied to the total mixture flowrate and to the liquid flowrate.

There is some confusion concerning relative errors because they are sometimes given in respect to each phase (for instance, error in oil flowrate measurement relative to oil flowrate) and sometimes in respect to the total mixture flowrate. Poor relative accuracy can be obtained with the first option when a phase is present at a very low fraction. The second option obviously leads to better results. The accuracy performance announced by vendors should be treated with a degree of precaution.

Similarly, relative errors are sometimes given relative to the "full scale" of the device and not to the measured values. For the vendor, this also has the advantage of offering better figures but this can be a source of misunderstanding because the "full scale" is not generally well-defined. The maximum flowrate of a phase generally depends on the simultaneous flowrates of the other two phases. Full scale relative errors are acceptable only if the full scale flowrate is well defined.

It is difficult to derive the accuracy of an MPFM from the accuracy of the basic measurements. In many cases, flowrate calculation involves a complicated combination of the various measurements. For this reason, MPFM accuracy is normally characterised by tests in multiphase flow. However, tests can also introduce various causes of errors (for instance, the fluid properties can be known with some uncertainty, the flow may not be well calibrated and may not be fully steady) and the reference measurements can have also their own accuracy.

At present, for multiphase meters, measurement errors of ± 5% of total flowrate and ± 10% of each individual phase flowrate are generally considered acceptable. Some operators specify an accuracy on gas rate, on liquid rate (for instance ± 10%) and water cut (for instance ± 5% absolute)

3.6.2.2 Repeatability

Repeatability characterises the ability of a measuring instrument to provide successive measured values close to each other when taken with the same instrument, at the same location and in the same flow conditions, but with different time intervals. This notion is broader than accuracy as a meter can have good repeatability and poor accuracy, but not the opposite.

For field operators which have to monitor the well development, this criterion can be more important that accuracy. A typical performance requirement for an MPFM is that repeatability discrepancies must be lower than 5% between consecutive tests on the same well, in the same conditions.

3.6.2.3 Sensitivity and Tolerance

Sensitivity and tolerance characterise the influence of quantities which can modify the measured quantities. For multiphase meters, sensitivity is related to fluid properties and flow disturbances. Pressure and temperature always have a significant impact on measurements and must always be measured simultaneously with gas and liquid flowrates, in order to be able to correct the measurement according to a common basis.

Other parameters can influence accuracy:

- density and viscosity of phases,
- water salinity,
- presence of sand, wax and scale contents, etc.
- change in flow pattern due to installation, range of operation.

3.6.3 Calibration

Periodic calibrations of MPFMs are required for two different reasons. The device response may need to be adjusted:

- to the true physical parameters of the produced fluids used in the metering principles, and after initial calibration to their long-term evolution,
- to the longt-term evolution of device response, produced by various disturbances, as in any metering system.

The basic measurement principles and the calculation of flowrates requires the input of certain physicals properties of the fluids, which may vary from field to field, but which may also vary over the long-term as the reservoir is depleted. Typical variable parameters are oil density (or °API), gas specific gravity, water density and salt content, absorption coefficients, dielectric constants, etc. Corrections are programmed for temperature and pressure variations but other unforeseen changes in these parameters introduce measurement errors which are unrelated to the quality of the device.

In service, the device response can also evolve over time due to a number of disturbing effects such as wear, wall erosion, solid depositions, corrosion, etc.

Generally, this calibration can be carried out by filling out the measurement section separately with each phase. The effect of velocity, if any, is not considered. Calibration is difficult in multiphase flow.

Direct comparison with separator measurements is not really suitable for calibration. As previously indicated, the accuracy of separators is limited and uncertain. It depends on the quality of separation which is never perfect, and the nature of fluids which may form an emulsion. Some liquid can be carried over in gas, and gas carried under in water or oil, oil in water, etc, leading to measurement errors. If this approach is however used, the quality of separation must be monitored.

Drift detection and the possibility of *in situ* re-calibration are of particular special interest in the case of possible solid depositions (scales, wax, asphaltenes). For this purpose, the installation arrangement must offer the ability to perform measurements with known single-phase fluids; gas, oil or water).

Steady multiphase flows can be used for calibration by recombination of single-phase flow.

3.6.4 Tests

Testing of multiphase meters includes functionality checks and comparison of measurement performance against a set of reference measurements (accuracy, repeatability and tolerance).

Several types of tests can be performed: factory tests, tests in multiphase loop facilities and field tests.

3.6.4.1 Factory Tests

Factory tests consist of functional tests and generally involve single-phase flow. They are often used to calibrate the flowmeter. Functional tests depend on the meter type and design. These tests are similar to tests of single-phase devices.

3.6.4.2 Tests in Multiphase Flow Facilities

The performance of a multiphase flowmeter must be tested to verify its accuracy and repeatability over the intended GVF, water cut and flowrate operating domain. To cover the full operating domain, a test programme requires a large number of measurements by comparison with single-phase meters.

Multiphase meters can be tested in realistic two-phase flow operating conditions in several facilities. Such facilities use "clean" fluids which can, for the liquid phases, be "dead" crude oil (after separation of gas) or synthetic mineral oil and water and, for the gas phase, air, nitrogen or commercial natural gas. A typical test loop consists of one or more separators to separate the basic fluids making up the mixture. Downstream of the separator, the single-phase fluids are metered and boosted separately and are then mixed. The mixture flows through the tested MPFM and then returns to the separator. Heat exchangers are used to stabilise the temperature of the mixture. Sometimes, when nitrogen is used, the gas is supplied by liquid storage gas tanks and is vented to atmosphere after separation. The loop is equipped with control valves to obtain the required flowrates. Additional equipment is also sometimes used to generate specific flow regimes such as slug flows. The MPFM measurements are compared with the measurements of the calibrated single-phase meters.

The accuracy of the multiphase loop must be checked because the significance of the reference measurements can be altered by the quality of separation during operation (accumulation of gas carry-under and liquid carry-over, same for oil in water and water in oil), or by mass transfer between phases after measurements (gas release when pressure drops), or flow instabilities in the test section. In general, a multiphase loop is unable to cover the full operating pressure and temperature domain of a commercial meter.

Multiphase flow loops of this type are available at the National Engineering Laboratory (NEL) in the UK and at IFP in France (Figure 3.11).

Figure 3.11

IFP multiphase flow loop at IFP-Lyon

3.6.4.3 Field Tests

Flow loops cannot fully reproduce the complex flow behaviour of wells or all the effects of live crude oil components. Only tests in true industrial operating conditions can demonstrate the suitability of an MPFM for an actual flow environment, its ease of operability and its ability to accept transient or dynamic behaviour, hardware robustness to extreme weather conditions, corrosive fluids, sand, scales, non-specialised operating staff, and so on.

Field tests are generally performed with the MPFM installed upstream of a separator. The meter measurements are compared with the measurements from the separator instrumentation. Instantaneous measurements performed with an MPFM usually show rapid fluctuations which are damped out in a separator. Comparisons therefore need to be made using average measurements over a long period, such as hourly averages.

However field tests have also their own limits:

- the accessible domain of flowrate, water cut, GVF, etc. is often limited and the full possible operating range of the device is not always covered.
- measurements must be compared with care (test separators are not very accurate, sometime less so than the flowmeter). In any case, the data must be corrected to a common P&T reference between the test separator and MPFM if this correction is not made automatically.

3.7 FIELD EXPERIENCE

3.7.1 Extensive Field Testing of MPFM

There are too many MPFM field applications to be reported thoroughly, with several hundred MPFM already having been installed on production sites around the world. This section therefore concentrates on a few selected examples with the aim of giving the reader information on the impact of successful MPFM installations.

3.7.1.1 Agar MPFM-400

Among the first meters that were seriously field-tested was the MPFM-400, a 2-inch intrusive meter manufactured by Agar Corporation [Mehdizadeh & Farchy, 1995; Tuss *et al.*, 1996; Ngai *et al.*, 1997]. The Agar MPFM-400 series was derived from an earlier meter from Agar (MPFM-300 series), to which a gas by-pass device was added (FFD^{TM}-Fluidic Flow Diverter) with aim of measuring high Gas Volume Fraction (GVF) flows.

Before field evaluation of the MPFM-400, the measurement performance of the MPFM-300 basic meter was also assessed by means of flow-loop and field tests. The test conducted on the flow facility in the Maraven S.A. Pilot Project at San Diego Norte (Orinoco Oil Belt – Venezuela) confirmed that the technology was suitable for measuring total flowrates in the 1 to 28 m^3/h range, with oil viscosity up to 2,000 cP (9.6 API Zuata heavy crude oil) and water salinity from 100 to 20,000 ppm. The relative measurement accuracy on the oil, water and gas flowrates was found to be always better than +/– 15% [Colmenares *et al.,* 1995]. It seems that no emulsion was observed in these low flow conditions.

The MPFM-400 was installed and performance-tested on the pilot test ground at Lafayette, Louisiana, operated by Conoco and on the Rockyford Battery producing field in Alberta, operated by PanCanadian.

The tests on the pilot loop were mainly aimed at assessing measurement performance with wells fluids. The reported relative measurement accuracy was within +/– 10% over the range of GVF from 0.2 to 1 and at static pressures of around 10 bar. A number of repeatability experiments were run successfully but no assessment of the measurement drift with time was reported.

During the field tests at Rockyford Battery [Ngai *et al.*, 1997], the meter was inserted on the flowline between the well test header and the well test-separator. Therefore the MPFM was measuring the flow immediately upstream of the test-separator that was used as a reference for flow measurement during the tests. This solution allows comparison of the measurement data with very small errors due to the thermodynamic correlation used for volume flows corrections and to the method used for measurements synchronisation.

Again, the MPFM-400 averaged measurements coincided well with the reference measurements (within +/– 10% on oil, water and gas flowrates) within the total volume flowrate range from 2 m^3/Hour to 6 m^3/Hour and thermodynamic conditions of about 10 bars, 25°C. In addition, the MPFM-400 was used for real time trending of gas, oil and water flowrates on this site. It helped to detect early GOR, WOR and slugging flow trends. This new information provided the operator with valuable insights into well behaviour, its artificial lift performance and its well bore flow conditions.

The MPFM-400 was also field tested in parallel with other MPFM by various companies [Mohamed *et al.,* 1998; Al Taweel & Barlow, 1999; Hasebe *et al.,* 2004; Bekkousha *et al.* 2004].

3.7.1.2 3-Phase Vx MPFM

Significant field testing work was also carried out on several production sites by the 3-Phase company, a subsidiary of Framo Engineering and Schlumberger that joined forces to develop and market the Vx MPFM.

After a series of calibration test periods run in several multiphase flow facilities (NEL, IFP-Lyon and others) within the total flowrate range at meter of between approximately 4,000 and 35,000 bbl/day, a 4-inch prototype of the Vx MPFM was sent for field testing in a variety of flow conditions [Atkinson *et al.,* 2000; Khoory *et al.*, 2003; Warren *et al.*, 2003; Hasebe *et al.,* 2004; Bekkousha *et al.* 2004].

This allowed upgrading of the viscosity correlations and the physical model used in the MPFM software for computation of flowrates in multiphase mixtures with heavy oil [Atkinson *et al.,* 2000]. At the PDVSA Intevep Experimental Test Production Centre in Venezuela, the Vx MPFM was tested with oil viscosities up to 3,000 cp and static pressure at the meter within the 10 to 30 bar range. With the adjusted model, the relative error on gas, oil and water flowrates was reported to be within +/– 10% of readings and repeatability was found to be very good.

The Vx MPFM was then sent to Venezuelan onshore fields to perform over 50 test jobs with a separator in series providing the reference measurements. Once again, the reported accuracy was within 10%. Moreover, the operators began to quantify the benefits that well testing with MPFM can offer with regard to ease of installation, operating safety, time and cost savings, etc.

In the Middle-East, the Vx MPFM has been successfully field tested onshore by ADCO [Khoory *et al.*, 2003] and offshore by Saudi-Aramco on a test-barge on the Safaniya field [Warren *et al.*, 2003]. In these widely differing conditions, the MPFM was always reported to be reliable, safe and easy to use.

On the Safaniya field, the Vx measurements were compared with reference test separator results. In Abu Dhabi, ADCO compared the Vx measurements with the results from a test separator and from downhole tool interpretation using the black oil model. Both tests concluded satisfactory measurement accuracy of the Vx MPFM in their particular conditions.

Furthermore, they also highlighted the great benefits of having continuous flowrate measurements during well tests, thanks to the MPFM, in order to acquire cumulative oil and water volume, production rate and operating conditions data (pressure and temperature) in real time without requiring separation of the fluids.

Multiphase flow metering can add considerable value during the exploration and development phase of a new field. Because MPFM can be installed in-line on the well stream and provide instantaneous flowrate measurements while applying very low back pressure on the well, they can be of great value in flow history quantification and decision-making during exploration and development well tests.

During the exploration and development phases of the Girassol field, deep offshore Angola, Total successfully used a Vx MPFM in addition to conventional well-test equipment [Mus *et al.*, 2002a & 2002b].

During the first exploration well tests at Girassol, with conventional equipment, emulsion and/or foaming was a problem. These flow conditions heavily affected the quality of the flowrate data measured at the separator because the emulsion and foaming could not be satisfactorily broken. Owing to the deepwater environment (1,250 to 1,400 m) the riser cooling effect was strong which made it impractical to supply the quantity of heat needed for emulsion breaking at the surface. Consequently, the flowrates measured were unreliable and good analysis of the potential of the wells was impossible with the conventional method.

The Vx MPFM technology was chosen for this application because it would not be affected by an emulsion or a foam and would be able to deliver correct gas, oil and water flowrates under these conditions. A Vx MPFM was designed for the working pressure rating of 345 bar needed for this application. This allowed the Vx MPFM to be installed immediately downstream of the choke valves manifold and upstream of the test separator during the drill stem tests and later development well tests.

Figure 3.12

Skid-mounted Vx (mark of Schlumberger) MPFM during tests at IFP (courtesy of Schlumberger)

With the Vx MPFM working within its design range of flowrates and GVF, there are many advantages over conventional procedures. Among the main benefits observed in this application were:

- elimination of safety valves and relief lines;
- fast and safe MPFM installation;
- instantaneous flowrate measurements;
- accurate measurements made at surface pressure;
- full record of flow periods and cumulative fluid volumes;
- accurate determination of well eruption;
- continuous water cut monitoring;
- very low backpressure on wells;

thus allowing:

- optimisation of well clean-up;
- optimisation of gas lift;
- better control on wellbore overdraw and sand-control completions;
- better knowledge of flared volumes;
- fast stabilization of production;
- increased personnel safety;
- cost and time savings;
- improved environmental protection.

3.7.1.3 Roxar 1900VI

The October field is located in the northern gulf of Suez in Egypt and has been operated by GUPCO since 1979 [Leggett *et al.*, 1996]. October produces a 27° API oil from the Nubia sandstone reservoir. The reservoir pressure had declined over the years and by 1986 gas lift was installed to maintain production.

In the mid-nineties October production was characterised by three-phase flow with GVF approximately 0.95 at surface conditions. Due to the October drilling development program and economic pressure on the company to maximise production on a daily basis, the surface flow lines and separation facilities were all used full-time to transport production. Consequently, well testing led to a significant decrease in production and the wells at October field could therefore only be tested once every 3 to 6 months.

Multiphase metering offered a solution for the frequent and high-quality well tests on the October field that were needed to manage the Nubia reservoir without deferring production. Other objectives were rapid identification of well problems and gas-lift optimisation.

The MPFM chosen for the October field was the non-intrusive Fluenta 1900VI (now Roxar 1900VI after the companies merged). The MPFM was installed on the production deck of a well-protector platform that was bridge-connected to the October production complex.

Individual wells were isolated from the production manifold, routed through the MPFM 1900VI, and placed on production to the test separator for a duration from 4 to 6 hours. The reported measurement accuracy as compared with the reference test-separator was within 10 percent in the GVF interval between 0.93 and 0.96 but deteriorated considerably on the liquid and the gas phase measurement for GVF > 0.96. However, this accuracy problem on the wells producing at very high GVF proved to be relatively unimportant for the purposes of managing the wells on the October field because the repeatability of the MPFM measurements was seen to be very good, thus allowing flowrate correction on the basis of previous tests data.

Finally, the MPFM installation on the October H platform proved to be an efficient alternative to the test-separator, given the objectives of this application, allied with minimum maintenance requirements and significant savings on test pipelines.

3.7.1.4 Esmer MPFM

The SJFT-C offshore platform installed in 2001, 35 km away from the Kota Belud shore in Malaysia, can be described as one of the first "smart" unmanned offshore platform in the world which can be controlled remotely through a well developed SCADA system. Production at the platform, operated by Sarawak Shell Berhad (SSB), has the characteristics of wet gas flow being in the region GVF > 95% with five wells producing at flow rates in the range 200-3,000 bpd, water cut less than 50% at an operating pressure of 10 bar.

A 3 in. Esmer MPFM from Petroleum Software Ltd was installed on the SFJT – C offshore platform in November 2002 [Cai *et al.*, 2004] and has remained in operation until March 2006 when it was replaced with a 2 in. Esmer MPFM. The replacement was needed by the drop in production of the gas lifting wells tied into the SFJT-C. The 3 in. meter comprising orifice with active shield capacitance technology have had a long trouble free stretch of use on the platform with no hardware faults encountered during the installation period.

Like all other instruments on the platform, the MFM was connected into the SCADA. Esmer measurements are downloaded to the Company mainframe through a MODBUS interface round the clock. SJFT-C is also equipped with a test separator which is used to validate and retune Esmer measurements at intervals.

The main objective of SSB in this instance was to benchmark the use of a multiphase meter on a remote unmanned platform and, depending on the outcome, to make wider use of MPFM on other platforms. Esmer was perceived to be a good first choice for benchmarking due to its emphasis on remote support, small footprint and straight-through flow design.

After three years of uninterrupted use, the Esmer experiment on this platform proved extremely positive from the viewpoint of reliability and ease of use (remote-access and calibration of the MPFM) as a means of unmanned well testing with significant cost savings for offshore operations.

In addition to its advantages for well testing, on-line measurement of GVF, water-cut and individual phase flowrates enables the operator to monitor changes from the data collection facilities continuously and provides a real-time view of the status of production and flow lines.

However, the accuracy of the measurements is still open to question across the broad range of flow conditions. The MPFM was shown to measure within the range of expected accuracy (5% liquid; 10% gas) provided that production is within the optimum region of the (two-dimensional) operating envelope. At SJFT-C, Esmer delivered the required accuracy for 2 out of 5 wells (producing over 1,000 bpd) but has not met the accuracy target in the lower production wells (producing under 400 bpd).

It is believed that turn-down ratio is the biggest practical problem to be overcome since the MPFM is often required by the operators to operate across a broad range of flows due to a combination of unknown production rates, mixed high/low producing wells and declining production. This range is sometimes bordering on the unrealistic for one individual meter.

Early in 2006, the decision to downsize Esmer to 2-inch, was taken in order to keep the differential pressure within the optimum 100 mbar to 1,000 mbar range and set the ends of the operating envelope at 20 mbar to 2,000 mbar. This works out to an approximate turn-down ratio of 15.

Six ESMER MPFM were supplied recently by Petroleum Software Ltd to White Nile Petroleum Oil Corporation (WNPOC) for use in the Thar Jath oil field in the south of Sudan.

Thar Jath is a new field development. The reserves of the Thar Jath oilfields have been estimated at a minimum of 250 million barrels. White Nile Petroleum Company is a consortium of Malaysian state oil firm Petronas, India's state-run Oil and Natural Gas Corp, and Sudan's state oil company Sudapet. The initial capacity of the Thar Jath field will be 80,000 barrels per day.

The operators perceived the main advantages of MPFM over separators as being lower capital and operational expense, ease of use, better production control and management through the SCADA interface and electronic reporting of the measurements.

Figure 3.13

Esmer V2 in WNPOC's Thar Jath oil field in Sudan (from PSL ltd)

Esmer systems were perceived to be superior to competing MPFM concepts due to the absence of radioactive source sensors, the simplicity and compact nature of the spool (straight horizontal pass-through of 1.2 m length without any flow obstruction), ease of use and remote-maintenance capability *via* the Internet.

The six MPFM supplied are identical in construction, all with 2-inch ID spools comprising a standard Venturi with Yokogawa sensors and a proprietary capacitance transmitter. The meters were tested at the NEL multiphase flow facility before delivery and demonstrated to be within the user's accuracy specification (5% liquid/10% gas relative accuracy and 3% water cut absolute accuracy).

The MPFM were installed at Thar Jath at the same time as the principal production facilities and were commissioned at the start of production during spring 2006.

Data processing is performed locally on an industrial PC/flow computer installed in the control room. This was recommended instead of the usual in-field flow computers in view of the hostile climate, which is hot and humid and prone to violent lightning and thunderstorms.

The objective of multiphase metering at Thar Jath can be described as well testing and production monitoring. There are six data collection stations at Thar Jath within a 5 km radius, and an MPFM was installed at each one of the stations. No test or production separators are available in the processing facilities. Thus the reliability and accuracy of the Esmer measurements will be critical for production management. At each data collection station, the MPFM is installed downstream of the test manifold feeding from up to 12 wells.

The oil has high viscosity (3,000 cp) with a tendency to form an emulsion and deposit wax. The wells are expected to produce a maximum of 3000 bpd per day, initially at low GVF (under 20%) and low water cut. Heat tracing tape was wrapped around the pipeline and the multiphase meters. Measurements taken by the MPFM are used by the operators for monitoring the relative performance of the wells.

Field commissioning followed four steps:

- Testing the original factory calibration.
- Adjustments for the physical properties of the produced fluids (mainly for viscosity, density, salinity, permittivity differences between laboratory and field fluids).
- Validation of the measurements against reference measurements obtained by other means (eg a manual Basic Sediment &Water (BSW) test is recommended as the first comparison followed by separator tests, where available).
- Any discrepancy which may be noted between Esmer and the reference measurement in step 3 is then corrected (akin to zero offset adjustment).

Zero offset adjustment is also scheduled at regular intervals to taking care of any tertiary effects arising from wax deposits, scale formation, blockages, etc., without the need for dismantling and clean-up operations. All stages of the work (calibration, validation, tuning) are performed remotely *via* the Internet as Esmer is specially designed to enable such remote-supervision. This brings the benefits of instant service for the operator and avoids difficult and costly journeys for the supplier.

3.7.2 Comparative Field Testing

As commercial MPFM are all different in terms of price, capability, strength and limitations, they cannot be procured easily through competitive bidding on the basis of a single set of technical specifications. Therefore, comparative testing of several MPFM on the same field is a good solution for evaluating the available meters against various criteria such as meter installation and commissioning, calibration and operation procedures, personnel training, after-sales services, maintenance requirements, HSE exposure, etc.

In quite different conditions, BP on North Slope in Alaska and Saudi Aramco on several fields in Saudi Arabia have conducted comparative field tests [Al Taweel & Barlow, 1999; Hasebe *et al.,* 2004; Bekkousha *et al.,* 2004]. Their conclusions agree that the measurements obtained from several meters tested on the same field often differ by more than +/– 10% although the vendors claim measurement accuracy better than 10%. Moreover, it is observed that the scatter in the measurement data obtained from several MPFM increases with the GVF. This result is consistent with the conclusions of MPFM calibration tests on multiphase flow-loops which showed that the measurement accuracy of the MPFM generally decreases at higher GVF.

Chapter 4

New challenges

4.1 INTRODUCTION

Multiphase technology applications have proven that they could bring substantial cost savings, especially in deep water and remote areas.

The successful experience with subsea rotating machines and flow monitoring have paved the way for the development of several types of subsea processing: re-injection of produced water after subsea separation, direct injection of raw sea water, subsea gas compression, etc. Several R&D programs are underway to extend the present state-of-the-art towards complex handling and processing of multiphase fluids.

In this chapter, R&D work in several areas is briefly presented and discussed: transportation of hydrates in slurry, subsea separation, subsea compression of wet gas and energy or recovery with a multiphase turbine. The requirements of oil and gas production in very deep water is a strong driver for these developments.

4.2 HYDRATE TRANSPORTATION IN SLURRY

4.2.1 Prevention of Hydrate Formation with Long Tie-Back

The risk of hydrate formation is a major subject of concern for oil production in deep oceanic waters. Gas hydrates are crystalline lattices formed in suitable temperature and pressure conditions by water molecules which enclose low weight hydrocarbon components (C1, C2…). Typically, hydrate formation occurs in production flow lines at high pressures and low temperatures such as those encountered in deep oceans [1] (0 to 4°C).

In a temperature – pressure diagram, the hydrate stability domain is limited by the dissociation curve (Figure 4.1). The location of the curve depends on the fluid composition. Some chemical components like salts, or methanol, glycol, etc. modify the hydrate stability

1. Natural gas hydrates are found in shallow layers of deep oceanic seafloor and in some gas deposits situated in permafrost soils in Arctic regions.

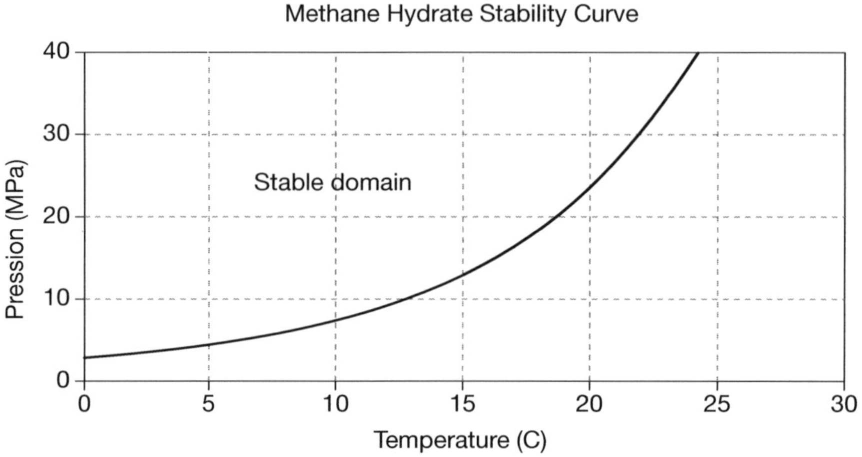

Figure 4.1

Typical hydrate dissociation curve

domain by moving the dissociation curve to the left of the diagram. They are called thermodynamic inhibitors. Others do not change the dissociation curve but prevent or delay the growth of crystals. They are called kinetics inhibitors.

Usually, in underground reservoir conditions, oil and gas fluids are warm enough to flow outside the hydrate stability domain. However, thermal exchanges with the surrounding water are sufficient to quickly cool the fluids in the hydrate domain after the well head when they flow on the sea floor. Pressure drops in flow lines and the gravity effect in risers produce gas expansion which contributes to cooling down the produced fluids. This combination of effects leads to a high risk of clogging of the flow line by formation of a hydrate plug. Once formed, hydrate plugs are difficult and dangerous to remove. They require lowering of the pressure, which is difficult in deep water, or heating of the line with a high risk of overpressure and explosion because hydrate dissociation produces a large volume of gas.

To be effective, conventional thermodynamic inhibitors such as methanol or glycol must be injected at high concentrations in respect to water content: something like 30% or more by weight. This option requires large regeneration facilities, storage capacity to compensate for losses and involves high operating costs. Moreover, a fraction of the inhibitor is dissolved by the oil which is depreciated since the inhibitor must then be removed by distillation.

This is why inhibitors are mainly used for gas production, when limited quantities of water are produced by condensation of the water vapour contained in the gas. For oil, current practice is to maintain sufficient fluid temperature by thermal insulation of well heads, flow lines and risers to prevent hydrate formation during normal production phases. Inhibitors are only used at start up of production when the flow lines are cold. At shutdown, the temperature gradually decreases with time. The gas and water contained in the lines are flushed and replaced by an inert fluid, usually a light gasoline, before entering the hydrate domain. In order to perform substitution of fluids, subsea flow lines must be designed with a looped configuration. The decision to flush a line is usually taken a certain time after production shutdown. For this reason, a minimum time is also required during which the fluid temperature must be kept above the critical temperature.

As the size of newly-discovered fields in deepwater basins decreases, stand-alone floating production systems are not economically viable. To produce new fields, subsea systems are now frequently tied-in to a remote facility. The facility can be either an existing one or a single new floating system which serves a cluster of several fields. Step-out distances exceed already 45 km for oil production (100 km for gas production) and longer distances are planned as at the time of writing.

The allowable temperature loss in a flow line depends essentially on the fluid characteristics (well head flowing temperature, hydrate dissociation temperature; etc.). The overall heat transfer coefficient U of the line (defined in section 1.3.6.4) must therefore decrease as its length increases. In other words, the thermal resistance R of the insulation must be increased proportionally to the step-out distance. An increase in insulation thickness quickly becomes inefficient as the thermal resistance depends of the logarithm of the thickness (as shown in section 1.3.6.4). The practical limits of thermal insulations capabilities are presently reached with U-values of 0.5 $W/m^2/K$. Long tie-back distances are limited by thermal insulation capabilities.

For this reason, active heating by hot fluids or electrical tracing has been used in the Gulf of Mexico in recent years in addition to insulation. However, this option is expensive in terms of energy and therefore in operating costs. Preference is given to protecting specific localized points such as a valve or a well head, or temporarily during transient operations such as shut-down or for warming a line at restart.

A new approach needs to be developed to enable a further increase in remote tie-backs of small deepwater fields. An ideal solution would be to allow the fluids to flow in the hydrate domain, at low temperatures, thereby eliminating the risk of blocking the line at the same time. It has for a long time been observed that hydrates do not necessarily block the line when they form. In case such as these, hydrate crystals do not agglomerate but are transported as a slurry of solid particles dispersed in the oil.

Two means are currently proposed to ensure stable dispersion of hydrate crystals:

- inhibition of crystal growth and agglomeration by injecting low dosage additives (LDA) into the multiphase mixture,
- formation of naturally anti-agglomerant crystals.

4.2.1.1 Low Dosage Inhibitors

Experimental results show that natural or artificial surfactants inhibit crystal hydrate growth. Dispersant additives are components soluble in oil. They act according to the following process:

- Formation of a water-in-oil emulsion by decreasing the water/oil interfacial tension.
- Prevention of coalescence of water droplets.
- Prevention of aggregation between hydrate particles.

The additives decrease the interfacial tension and prevent coalescence between water droplets. When hydrate formation occurs, water droplets are transformed into hydrate particles which are dispersed in the hydrocarbon liquid phase. Additives act as dispersants by preventing the formation of large aggregates.

IFP has developed a process which aims to transport hydrate crystals as a suspension in well streams. This process is based on the use of surfactants which are likely to create a water emulsion in liquid hydrocarbons (condensate or crude). The size of the hydrates particles is thus controlled and there is no agglomeration of crystals.

Dispersant additives are sometimes called "emulsifiers" or "anti-agglomerators". Table 4.1 gives a few characteristics of dispersant additives.

Table 4.1

Dispersant additives	Added weight %
EMULFIP 102 B	1.0%
IPE 201	0.5%

The additives do have some limitations. The maximum water cut is expected to be around 50% to obtain a first emulsion of water-in-oil. This limitation is influenced by the rheological properties of fluid emulsions and solid suspensions. It may depend on flow regime conditions.

4.2.1.2 Formation of Stable Non-Agglomerant Hydrates

The idea of forming naturally anti-agglomerant hydrate crystals with a specific formation process was proposed following observations by researchers at the SINTEF laboratory. The formation process is based on a rapid transformation of free water in non-agglomerated crystal hydrates by mixing the fluid mixture with a re-circulated fraction of the cooled fluid. The re-circulated fluids already contain many small hydrate particles which act as seeds to eliminate all water. This process is generally called "Cold Flow". It is patented and is currently under development.

4.2.2 Main Issues of Slurry Transportation

Transportation of hydrate in slurry raises four main issues: the ability to transport the slurry effectively, disposal or dissociation of hydrates at the host facility and cost.

4.2.2.1 Behaviour of Hydrate Slurries

The design of an efficient system to transport hydrates must allow prediction of pressure drops with enough accuracy to understand and model the rheological behaviour of this type of slurry.

Few models or experimental data have been published to date. Experimental studies of hydrate slurries are complex owing to the temperature and pressure conditions. The hazards involved in handling natural gas also require careful test procedures. Nuland and Vilagines [2001] presented a review of a number of slurry models. Peysson *et al.* [2003, a & b] published interpretations of tests performed in a special multiphase flow loop built at IFP-Lyon to work under controlled thermodynamic conditions (temperature controlled between 0°C and 50°C, pressure adjusted between 0.1 MPa to 10 MPa). Figure 4.2 shows a picture of the loop. The flow rates of the gas and liquid phases can be adjusted.

Figure 4.2

Thermally-controlled experimental flow loop to study hydrate formation and transport at IFP-Lyon

The authors investigated the behaviour of hydrate slurries up to 30% initial water cut, with mean flow velocities up to 3 m/s. Details of experimental conditions and operating procedures can be found in the referenced papers. For high flow velocities, in the turbulent domain, they found that pressure losses are well described by the usual square law model, but with an increased friction factor compared with the usual friction factor given by the Moody diagram or the Colebrook correlation. According to the authors, the friction factor can be written as the sum of two terms $f = f_1(k) + f_2(h)$, the first being the conventional factor, a function of the relative roughness of the pipe $k = \varepsilon/D$, and the second being an additional term which characterises the impact of the slurry flow and depends on the hydrate fraction.

For low velocities, corresponding to laminar flows, the interpretation of results is less clear. Basically, it has not yet been established whether the behaviour of the mixture is non-Newtonian or remains Newtonian under the influence of the particle accumulation in the lower part of the pipe. In this case, the overall behaviour would result from a complex mechanism with non-uniform distribution of hydrate particles in the cross-section. The downward settling of particles through gravity is balanced by shear diffusion with local apparent viscosity also influenced by the distribution of hydrate particles.

Further investigation is still needed on particle segregation in flow, as well as on many other subjects, such as hydrate particle volume fractions and behaviour at shut-in restart.

4.2.2.2 Other Issues

Obviously, there is an upper limit to the water cut for which hydrates can be transported in slurry. The hydrate quantity must be low enough for it to be able to flow as solid particles in the oil. Excess water must be removed in the case of production at high water cut. Subsea separation of produced water is presented in the following section.

The hydrates must be dissociated at arrival by heating or by chemical additives if this is not done naturally under pressure and temperature action. In addition, the compatibility of the various additives with the downstream process must be checked, along with the possible formation of foam or the tendency to form a stable emulsion.

Although injected in small quantities, low dosage inhibitors entail a price which leads to increased operating costs which must be compared with the lower investment costs.

4.3 SUBSEA SEPARATION

4.3.1 Experience of Subsea Separation

4.3.1.1 Review of Past Attempts

Subsea separation of well stream mixtures was considered and looked at relatively early on in the history of offshore oil, well before multiphase boosting or metering. However, it is not yet applied on a full-scale commercial basis. Phase separation often implies operational difficulties such as accumulation of sand, formation of an emulsion, foaming, deposition of wax. It requires a complex system with many active components to monitor and control the process which may require intervention and maintenance.

The first field trial of subsea separation was carried out in shallow water by BP and Total in the Zakum field (Abu Dhabi) between 1969 and 1972. (22 m) The trial included testing of a subsea well, of two successive separators and various other subsea components and control devices. The pneumatic control system, which was the conventional technology at that time, was found to be too unreliable for subsea operations [Goodfellow *et al.*, 1974].

Between 1974 and 1978, Exxon ran a pilot test of the Submerged Production System (SPS) in the Gulf of Mexico. The prototype tested comprised various types of equipment needed for subsea production, including a separation and pumping module. Unlike the Zakum test, the SPS was designed for diverless installation and maintenance, although tested at a depth of 52 m. All the subsystems were installed successfully and worked as planned, except for the separation and pumping module, due to a problem in the electrical distribution system [Burkhardt & Michie, 1979].

In 1988 and 1989, another field trial was performed in the North Sea, in the Argyll field operated by Hamilton Brothers. Argyll was the first field developed by a floating production system, in 80 m of water (1978). The prototype station was designed and developed by BOET (British Offshore Engineering Technology) and consisted of three horizontal vessels: a high-pressure (gas – liquid) separator placed above low-pressure separation (3-phase separation with a sand trap) and a storage vessel. The station also included a pump to export the liquid phases, pressure-controlled valves and a full set of pressure and level sensors. The design was in fact directly derived from conventional surface units, and was relatively complex in term of process and instrumentation. Separation performance met the objectives, but many reliability problems were encountered [Songhurst & Edwards, 1989].

Several other projects were developed and tested onshore: Deepsep, KBS, GA-SP, GLASS, VASPS, Alpha Prime. Among them, the KBS (Kværner Booster Station) combined a small separator with a pump and a compressor [Cotton and Stinessen, 1994]. This development could be considered the precursor of the wet gas compressor stations addressed below (subsea gas compression section).

The GA-SP project (1989) introduced several new ideas for surface separation, such as a modular approach for easier maintenance, liquid level monitoring by nucleonic sensors and electrical instrumentation. Control of valves was still hydraulic. The system worked well, but problems were encountered with wet electrical connections. Alpha Prime carried on the GA-SP approach with an all-electric system under the Æsop project funded by the European Union.

In all these projects, separation was essentially gas-liquid separation (gas and oil or gas and oil-water mixture) sometimes with three-phase separation. Downstream, production was exported by single-phase pipelines and, when required, boosted by a single-phase pump and compressors. None of these projects, except for the VASPS (Vertical Annular Separation and Pumping Station), have yet resulted in full production operations. However gas-liquid separation is being considered for a number of development projects.

Attention returned to subsea separation few years ago, with the idea of limiting the separation to bulk water. In this approach, the produced water is directly re-injected in a disposal well without passing by a surface facility. Oil and gas flow to the surface in a multiphase mixture, with little residual produced water, for separation and processing. This concept has a number of advantages, detailed below, especially in deep waters. In 2000, a first subsea experimental water separation station was installed in the Troll field (North Sea).

Several systems designed for downhole separation of produced water have also been developed: Aquanot (by Schlumberger-Reda), Hydrosep (by Baker Hugues-Centrilift), etc. They are implemented using well completion technologies, associating hydro-cyclones and electro-submersible pumps (ESP). For the time being, the flow rate capacity of these systems is relatively limited when compared with the productivity of modern subsea wells and these systems aim primarily to improve the activation of onshore wells.

4.3.1.2 Gas Liquid Separation: the VASPS

The VASPS is an innovative concept invented by the BP labs and developed by Agip, Mobil and Petrobras. In contrast with the other projects, the VASPS does not use gravity separation in a conventional vessel but centrifugal acceleration in a helical channel. The helical channel is installed in the seabed in a 30-inch conductor pipe with an overall height of between 30 and 80 m. The gas flows at the upper part of the helical channel. At the bottom, a conventional ESP pump boosts the liquid phase to the surface facility. All the installation and maintenance operations are performed by conventional drilling and completion technologies.

A prototype has already been tested on an onshore field in Italy then subsea in the Marimba field, in 395 m of water offshore Brazil in 2001 [do Vale *et al.*, 2002]. After failure of the pump due to a valve which remained closed during a restart operation, the control system was improved to avoid ESP overloads and the experiment was resumed. Since this modification, and as at the time of writing, the VASPS system has been in continuous operation for 2 years and 5 months.

4.3.1.3 Produced Water Separation: the Troll Pilot Station

A first experimental water separation station was installed in 2000 in the North Sea, at the Troll field operated by Norsk Hydro. The Troll Pilot is located in 340 m of water, 3.5 km away from the Troll C platform which receives production. The processing capacity of the station is 6,000 m^3/d of water, 4,000 m^3/d of oil and 800,000 m^3/d of gas. The separator has a relatively simple design with an inlet device to avoid formation of an emulsion, an oil-gas mixing system at the outlet and a sand jetting system. Externally, it has thermal insulation to prevent hydrate formation in the case of shut-down. The water stream is directly re-injected into a well located on the same template with a centrifugal pump. The pump is driven by a 2 MW electrical motor. The oil-water level is controlled by pump speed variations [Horn *et al.*, 2003].

After initial problems with the high-voltage electrical connector, the Troll Pilot worked satisfactorily but separation efficiency was slightly below initial objectives with 200 ppm of oil in water. Nevertheless, direct re-injection of water reduced the flowrate processed on the Troll C platform by some 8%.

4.3.2 Subsea Separation in Deep Water

4.3.2.1 Introduction

Offshore fields are now being developed in deep and ultra-deep waters, down to 2,700 m and more below the sea surface. Subsea separation is particularly attractive at very great depths, where it can be part of an anti-hydrate policy. As already mentioned, the conditions encountered in deep water are typically within the hydrate formation domain.

If separation performance is sufficient and residual water content in oil is limited to a low water cut (around 5% for instance), anti-hydrate treatment is no longer required or is limited to an injection of low dosage additives. Water separation near the well heads also offers some other advantages detailed below.

The Troll Pilot station demonstrated that reliability problems have now been overcome, but a number of issues still surround the application of separation in deep water. Several projects to develop the required technology are on-going.

4.3.2.2 Advantages of Water Separation in Deep Water

Subsea oil-water separation offers a number of benefits in deep water. The main ones come from simplification of the hydrate prevention policy, easier oil-water separation, improved well productivity, downsizing of water treatment in surface facilities and less produced water disposal at sea [Alary *et al.*, 2000].

- **Simplification of the Hydrate Prevention Policy**

If the residual water cut is low enough after separation, no anti-hydrate treatment is required if the natural dispersing properties of the crude are sufficient. The use of low dosage additives (LDA) could be an alternative option (described in section 4.1: slurry transportation of hydrates). This simplification offers several important advantages, due to reduced flow line insulation requirements and to simplification of network complexity.

When the risk of hydrate formation is the governing parameter in the design of the insulation system, *i.e.* for a non-waxy oil field, subsea separation could obviate the need for any insulation. All the problems associated with under-coating corrosion and its impact on cathodic protection levels are eliminated. All problems associated with large line apparent weight and on-bottom stability are considerably simplified. The savings associated with simplifying the double insulated pipeline can to a large extent compensate the cost associated with the subsea separation system.

On deep offshore developments, subsea wells are generally connected by a pair of subsea flow lines allowing pigging, flushing, scrapping and depressurisation from both sides using a loop network (called a "daisy loop" arrangement). However for low-cost field development or fields on which wax deposition is not an issue, a single line can be installed, saving on flow line costs and installation costs. In that case, present hydrate prevention strategies are no longer applicable and subsea separation should be considered. In addition to simplicity, the savings should also take into account the additional advantage of subsea separation on insulation requirements.

- **Improved Well Productivity**

The improvement of well productivity by lowering the water cut is obtained by reducing the total flowrate in flow lines and by reducing the average density of the multiphase mixture. Both effects contribute to lowering the back-pressure at the wellhead. The gain in pressure is significant in deep water. The apparent viscosity of oil-water mixtures increases significantly with the watercut. The reduction of viscosity with water separation also contributes to reducing pressure losses.

The reduction in Wellhead Flowing Pressure (WHFP) leads to an increase in well flowrate or a reduction in the boost pressure or gas lift injection, if they are required. This can lead to a lower abandonment pressure and therefore an increase in the recoverable reserve.

- **Other Advantages**

The oil is generally less viscous at wellhead than at surface due to higher temperature and pressure. The direct influence of temperature and pressure reduces the viscosity. The greater amount of gas dissolved in the oil also contributes to reducing the viscosity. Compared to surface conditions, the reduced fluid viscosity leads to an improvement in separation efficiency.

Subsea water separation also offers a number of other benefits: possibility of phased development, reduction in size, weight and the associated cost of the production water treatment unit in the surface facilities, reduction in production water discharge to the sea and consequently oil disposal to the sea, which can only be environmentally beneficial, even though the oil content of discharged water meets regulation criteria.

4.3.2.3 Advantages of Gas-Liquid Separation in Deep Water

Subsea gas-liquid separation still has some advantages, despite the development of multiphase technologies. The main benefits put forward are reduced gas compression, with simplification leading to capital and operating cost savings, and simplification of the prevention of hydrate formation.

- Gas compression: the gas can be flowed directly to surface by a dedicated line at relatively low pressure. Gas compression can therefore be eliminated or considerably reduced. This implies a significant reduction in energy supply and simplification of the required subsea rotating machines. Less expensive single-phase liquid pumps can be used if separation performance is sufficient.
- Simplification of the hydrate prevention policy: less need for glycol (only for gas resulting from degassing in the oil line), less insulation.

4.3.2.4 Main Issues of Subsea Separation

Several issues must be resolved when designing a subsea separation module. They have a major impact on equipment selection and play an increased role directly linked to the greater water depths when compared with the existing solutions for shallower water.

- **Separation Performance**

Separation efficiency requirements are specific to each application. They may be influenced by the domain of hydrate formation, the export conditions and tie-in distance for the oil stream, and by the reservoir characteristics for the produced water stream. They must be satisfied across wide variations of flow characteristics.

Typical specifications for separation performance are approximately as follows:

- oil stream: maximum water content in oil: 2% to 10% (depending on oil composition, temperature, export distance and injection of additives)
- produced water stream: maximum oil in water: 50 to 500 ppm
  maximum sand content: 1 to 10 ppm
  (depending on formation characteristics)

Excessive levels of oil or solid particles in the re-injected water can reduce the formation permeability and reduce the injection capacity.

- **Sand Management and Disposal**

Production of sand can be significant in some fields. Moreover, water separation or multiphase boosting will probably tend to increase sand production, as well productivity is increased.

Two possible problems are associated with sand: accumulation in dead volumes and excess flow in the re-injected water stream, which can obstruct the reservoir pores.

The sand can settle and accumulate in low-velocity areas, such as the separator tank. It can also be a source of erosion for rotating components, valves, chokes, etc. Means must be provided for detecting and managing potential sand problems.

A system to monitor the presence of sand in the well stream and its accumulation in the separator must be installed. Acoustic measurements are already used to monitor sand on topsides.

- **Interface Level in Separator**

Reliable monitoring of the interface levels in the separator must be provided to control the process. The presence of foam and emulsion must be detected. Nucleonic detectors are considered for this purpose.

- **In-Line Measurement of Oil Content in Re-injected Water**

Excessive concentrations of oil in the re-injected water can reduce formation permeability and reduce the injection capacity. In-line monitoring and control of the residual oil concentration in water in the range of low concentration will be necessary. At present, on the Troll Pilot, only discontinuous monitoring is performed by sampling the produced water with a ROV. The samples are analysed using laboratory methods. This operation is a complicated one, especially for future systems installed in deep water.

- **Pressure Design**

The external pressure is the hydrostatic pressure of the seawater surrounding the pump module. During operation the separator is generally operating below the hydrostatic pressure in very deep water, but variable pressure conditions could be experienced in the separator and in the pump. All pressure parts of the module shall be designed for the full external hydrostatic pressure, considering that the riser may be filled with gas and depressurised to atmospheric pressure. They shall also withstand an internal pressure corresponding to a Wellhead Shut In Pressure (WHSIP). In very deep water, the required wall thickness can lead to very heavy pressure vessels and piping. R&D is underway to replace steel by composite materials in order to reduce their weight.

- **Reliability**

The obvious gains of using such a system depend on the reliability and maturity of the subsystems concerned. Seabed separation modules have already been installed and operated at shallow to intermediate water depths. In addition, the control and operation of the module is not very different from that of a complex subsea manifold or wellhead already installed. Finally, as for the pumps and associated power supply, their reliability is now felt to be proven.

Interventions, however, are inevitable and they must be planned at the design stage. Most of them should preferably be performed with ROVs, which have limited ability to lift components or apply loads for operating valves, and preferably without any need for depressurising the subsea system.

- **Flow Hydraulic Behaviour**

A potential problem associated with subsea separation is the impact of reduced liquid flowrate on the flow behaviour. Reducing the liquid flow rate in a line may induce hydrodynamic slugging or severe slugging in the riser. This problem must be evaluated on a case by case basis, as there are solutions such as gas-lift in the riser or top-riser pressure throttling.

- **Pump Duties for Water Injection**

Since the separator is operating below the required injection pressure, large differential pressures are required for the water re-injection pumps. Furthermore, generally high flowrates must be processed. Both flow and pressure requirements generally result in installed power levels for re-injection in the 1 to 2 MW range and even sometimes higher. Process fluctuations produce load changes in rotating equipment which must be accommodated in some way. In the case of a re-injection pump driven by an electric motor, this is generally solved by speed regulation with a variable-speed system.

- **Separator Design**

Conventional separators are generally proposed. Their high internal volume acts as a buffer tank and they can accept relatively large flowrate variations. Other types of compact separation systems are often proposed instead or in addition to improve the efficiency of separation, such as cyclones, electrocoalescers, floatation.

4.3.2.5 DIPSIS: a Typical Water Separation Station

Several deepwater separation stations have been developed and are proposed. Figure 4.3 shows a typical example of a subsea station. This project, called DIPSIS, is developed by a group of companies under the leadership of Doris Engineering: Prosernat, Sulzer Pumps, Cybernetix, IFP and ECA. The station was studied and designed to operate at mud-line level in 1,500 m or more of water [Alary *et al.*, 2002]. It comprises five modules:

- Inlet and outlet manifolds.
- Separator.
- Pumps manifold and control system.
- Two re-injection pump modules.

Each module can be retrieved and re-installed independently. They are clamped to a template designed to connect them together hydraulically and electrically.

Each module is designed to be easily accessible by ROV for retrieval of critical items for maintenance.

The design of such a station must account for the environmental conditions of deep offshore units: high external pressure, modular installation and retrieval, ROV access and interfaces, high-reliability through redundancy of critical systems, modular subsea control system.

A gravity separator was selected in order to obtain the most reliable design and a large available turndown without modification of internals. The internals are selected in order to

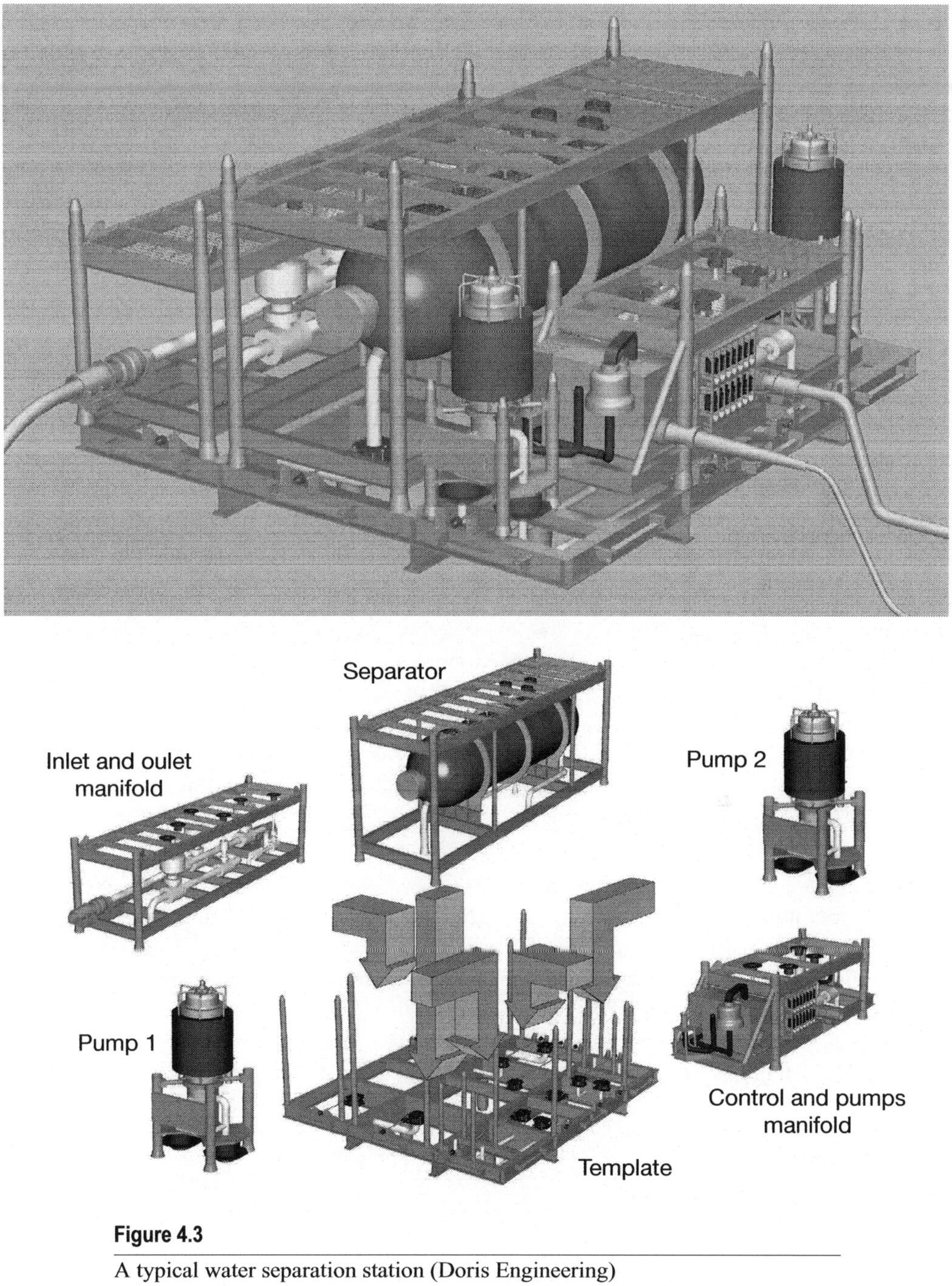

Figure 4.3

A typical water separation station (Doris Engineering)

reduce the risk of blockage. Oil/water separation is performed by gravity in the horizontal section of the separator. In deep water, the separation pressure is high and the free gas fraction lower than on the surface, allowing a larger volume for oil and water separation.

All pressure and temperature instruments are redundant. Level instruments are duplicated and use two different technologies. The control system is modular and can be easily upgraded. The oil/water interface level is measured by gamma-ray detectors and regulated by the water injection pump speed. Gamma-ray detectors give a complete cross-section profile inside the separator: sand, water, emulsion and oil levels are detected. Gamma-ray detectors are non-intrusive and can be replaced while the separator is in operation. They are unaffected by deposition and have no moving parts. A back-up level measurement system is added to increase the availability of the overall system. A specific device based on microwave measurement is currently under development. If sand accumulation is shown by level monitoring, it can be removed through a sand jetting facility inside the separator. High-pressure water injection will evacuate the sand downstream of the separator and through the trunk-line up to the surface.

With overall horizontal dimensions of 12×10 m for the template, the separator has an external diameter of 2.46 m and a length of 7 m. These sizes enable a processing capacity of up to 45000 bbl per day of liquid, covering the production of a cluster of three to four wells, with separation efficiency better than 10% Vol. of water in oil and 0.1% Vol. of oil in water, with oil densities ranging from 825 to 920 kg/m^3 (°API: 22 to 40). The maximum injection capacity is 80% of the design liquid flowrate or 36,000 bbl per day. The station has an apparent submerged weight of less than 200 metric tonnes and can be installed in a single piece.

In order to increase the availability of the module, two pumps are installed (*i.e.* one running and one on standby) and are powered by a single electrical cable. Hydraulic actuators are used to operate the power switch system to select the second pump. Each pump is completely and independently retrievable. The power is supplied from the topsides main distribution system through a step-down transformer, the variable-frequency drive system and a step-up transformer to the subsea electrical motor at 6.6 kV [de Salis, 2000].

A systematic approach to pump module design with functional analysis has been adopted to ensure reliability. The emphasis has been on simplicity of the design, installation and operating procedures. The design also incorporates as many already proven sub-components as possible. The key objectives are to minimise the risks inherent to greater depths and to focus on a high degree of system reliability.

The thermal insulation of the separator is designed to maintain the temperature 6°C above the hydrate dissociation curve for shutdown periods of up to 8 hours.

4.3.3 Conclusion

Subsea separation has now reached the stage of successful field testing and is on the way to full-scale operational applications. This has been made possible by a move towards simple one-stage separation systems with all-electric instruments and control systems. The instrumentation of future systems will not simply deal with the interfaces but will also probably be able to detect fluid-related operational problems, such as the formation of foam and emulsion and sand accumulation. Subsea control valves are eliminated. Pressure is controlled from the surface and liquid levels in the separator by variable-speed pumps.

4.4 SUBSEA GAS COMPRESSION

4.4.1 Introduction

Subsea compression provides the ability to transport gas over very long distances without the need for surface facilities and to maintain production capacity when field pressure declines. The main benefit of subsea gas compression is a significant reduction in costs by eliminating surface facilities in deep and very deep waters.

Subsea gas compression is also of special interest in Arctic seas, where large gas fields have been discovered. Subsea developments are the only viable production systems since the presence of ice packs or icebergs, at least for several months of the year, prevents the permanent use of floating systems.

Several subsea gas fields in the Gulf of Mexico, the Norwegian Sea, the Mediterranean sea (Figure 4.4) are currently produced directly from the wells to the shore or a remote offshore platform through long-distance pipelines. Step-out distances at present reach 110 km. Reservoir pressure is today sufficient for gas transport, but subsea compression may be of interest for all these fields after some 10 to 15 years of depletion.

In the cold conditions of deep waters, a natural gas is generally "wet" and this is one of main issues for subsea conditions. A wet gas is a natural gas which contains enough heavy hydrocarbon components to form a liquid hydrocarbon phase when it flows from the reservoir to the production facilities. Some liquid is formed when the dew point line of the phase envelope is crossed as the fluid pressure and temperature drops (Figure 4.5, see also Figure 1.19). The liquid produced is a light hydrocarbon phase (generally above 50 degrees API) called a condensate. A natural gas is said to be "dry", as opposed to "wet", when its composition is such that the fluid remains entirely in a gaseous state.

According to the above definition and strictly speaking, the terms wet and dry refer only to the production of a liquid hydrocarbon phase and not to the moisture content. However, a natural gas can contain some water vapour if it is in contact with liquid water in the reservoir. The gas water content corresponds to the saturated water content in reservoir conditions. Therefore, even a dry gas can produce some liquid when its temperature drops and water vapour condenses. In the rest of this text, this situation is also included under the term "wet" gas in the discussed topics.

Due to the considerable difference in density between the gas and liquid phases, conventional compressors cannot accept liquid droplets with volumetric fraction in excess of about 0.5% without damage.

Several compressor concepts are currently under development, but subsea compression is complex and many technological issues have to be resolved. High power consumption is generally required in gas fields, usually ranging from 10 to 50 MW.

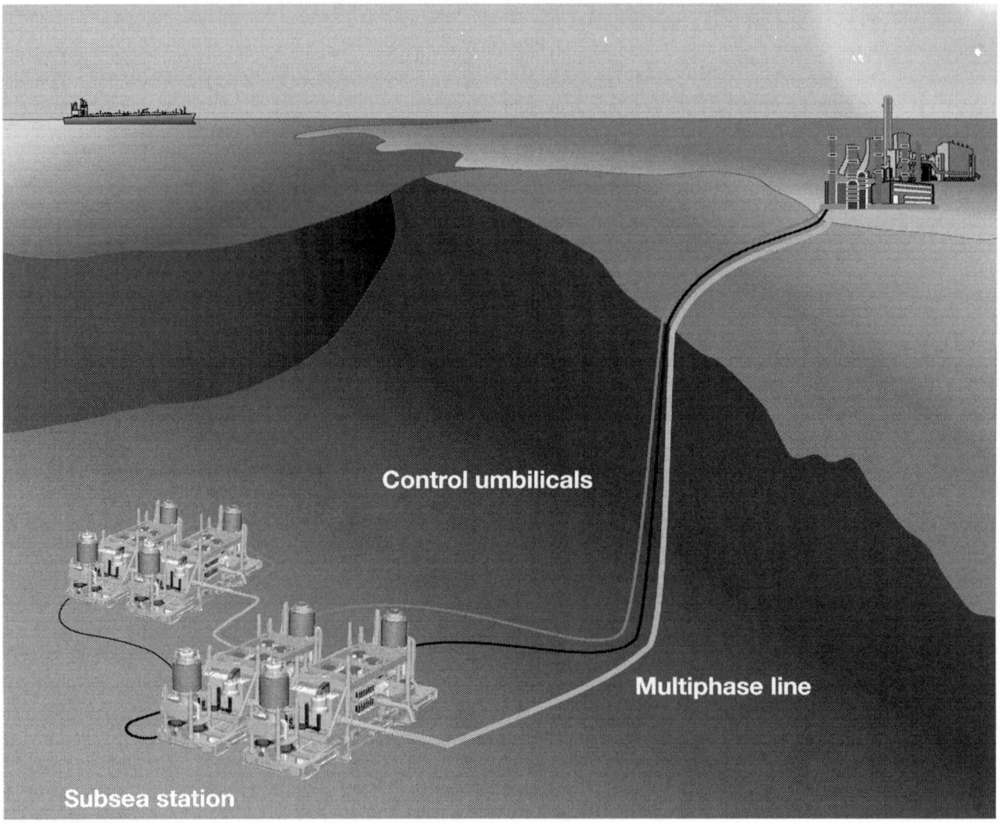

Figure 4.4

Typical subsea deepwater development

4.4.2 Technological Concepts

4.4.2.1 Compression Systems

Two main approaches are used by the developers of subsea compression systems:

- a compression machine design that can accept some liquids in the gas. This approach can be seen as the next logical step beyond multiphase boosting;
- a system design which includes the separation of liquid, a compressor for the gas stream, boosting of the liquid stream and remixing of liquid and gas.

4.4.2.2 Direct Compression of Wet Gas

Early applications of multiphase boosting started with GVF lower than 50% (volume throughput of the liquid higher than that of the gas), although this value was then rapidly

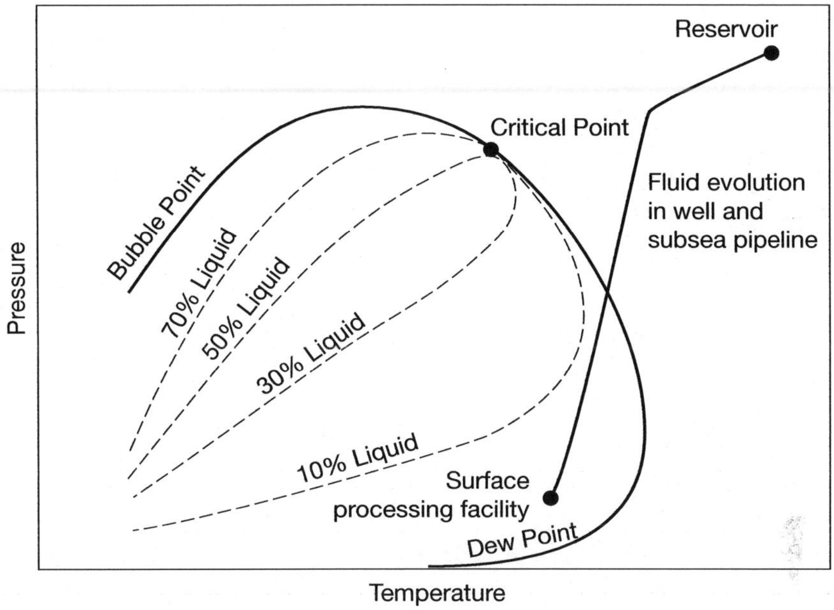

Figure 4.5

Wet gas

exceeded. Nowadays, many operators wish to transport a two-phase mixture with a volume of gas which is considerably higher than the volume of liquid. In these cases, ratios of gas to liquid volumetric flowrates are in the order of 100 or above.

Multiphase boosting of oil and gas mixtures at GVF exceeding 98% continuously is currently performed with twin-screw MPP for steam recovery of heavy oil in several onshore Canadian fields. The discharge pressure is relatively low, but the compression ratio is generally high, at above 6-7. For these applications, temperature rise becomes an important issue [Müller-Link *et al.*, 2002]. Shaft sealing, oil droplet vaporisation and steam condensation are other technical issues to be considered. For the anticipated subsea applications of wet gas compression, the compression ratio is lower – about 2 – but with high flowrates, higher than those achieved with steam recovery. In order to increase the efficiency of gas compression, a new screw design, with variable pitch, is being developed and tested by J.H. Bornemann, a leading manufacturer of multiphase pumps [Müller-Link *et al.*, 2004].

Framo Engineering has developed and tested a subsea rotodynamic wet gas compressor. When the GLR is about 100, the density of the mixture is relatively close to that of the gas and the compression ratio can only be achieved by using relatively high peripheral velocities. To limit the rotation speed the Framo Engineering compressor is based on the contra-rotating concept. This comprises several axial impellers assembled in two intermingled series which are rotating in two opposite directions at 4,500 rpm. The relative rotation speed seen by the gas is thus 9,000 rpm. The first series is assembled on a central shaft and the second on a peripheral casing. The two series are driven by two independent electrical motors placed at each end of the compressor.

Several prototypes have been tested by Framo Engineering. The last one – the WGC 2,000 prototype – has 20 impellers and two 1.8 MW electrical motors, providing a total output of 3.6 MW. At a suction pressure of 70 bar, the wet gas compressor with a capacity 3,000 m^3/h can deliver a 30 bar pressure increase. The compressor can accept 100% gas as well as 100% liquid and is sand-tolerant.

This prototype was tested at the company facility near Bergen, in spring 2003, then with live gas and condensate in the Statoil wet gas test facility at Kårstø in Norway. A design for surface applications is also available.

4.4.2.3 Subsea Compression of Dry Gas

A dry gas compressor is more efficient than a wet gas machine and this may be of significant interest considering the power needs involved in gas compression. A dry gas compressor however typically requires an efficient subsea liquid separation system upstream of the compressor unit.

As part of their "Subsea Centrifugal Compressor Module" (SCCM) project, AkerKvaerner and General Electric (formerly Nuovo Pignone) designed a subsea dry gas centrifugal compressor line, called Blue-C, with units of 2.5 MW and 5 MW. An 850 kW, 6-stage prototype has been tested onshore [Tasselli *et al.,* 2003]. This prototype was a refurbishment of the compressor module of the Kvaerner Booster Station project (KBS) designed in 1990-1992 and tested in 1993 [Cotton and Stinessen, 1994]. The new module includes an improved gas sealing system and an anti-surge re-circulation valve operated by a fast hydraulic actuator.

Another subsea centrifugal compressor project is being developed by FMC Technologies and Siemens Power Generation. The compressor concept (EcoII) is derived from a totally enclosed process gas compressor built for the Groningen gas field in the Netherlands.

4.4.3 Technical Issues of Subsea Compression

All these concepts are however in the early qualification phase, concerning maintenance, reliability and availability. Several other important issues are being investigated by the developers.

4.4.3.1 Compressor Design

Centrifugal compressors generally involve high-velocity rotating parts. Impact by sand particles and liquid droplets, even small ones, quickly leads to blade erosion. Solid and liquid particles must be eliminated

Many sub-systems are associated with a compressors for monitoring and protecting the machine. They include:

- process cooling,
- anti-surge control,

- gas-liquid separation (safe operation including control to accommodate slugs in transient operations),
- liquid handling around compressor,
- lubrication,
- gas-cooled magnetic bearing (no seals) with magnetic bearing control at distance.
- improved reliability: MTBF of centrifugal compressor on topside is at present somewhat low for a subsea application.

They must be designed for subsea operation and remote control.

4.4.3.2 Electrical Supply

Compression on a gas field generally requires high power consumption. Power needs can reach 50 MW. This is relatively high power when compared with MPP (something like a factor of 10 or more for a field development). Long supply distances will be needed. The present tie-back distance is around 100 km but it is felt that it could be up to 500 km.

This requires development of full distribution and remote control of subsea power.

4.5 MULTIPHASE FLOW TURBINES

4.5.1 Introduction

Recovery of hydraulic energy from a two-phase flow may have an economical interest.

Several types of applications have been identified to date [Charron *et al.*, 2003, 2004] in oil production as well as in other industrial processes. They are briefly presented below.

The use of two-phase turbines offers similar advantages to those of single-phase turbines except that they are designed to accommodate two-phase flows.

4.5.2 Typical Applications of Multiphase Turbines

4.5.2.1 General

Multiphase turbines can be used in two main categories of processes:

- for the replacement of a choke or let-down valve in two-phase flow,
- in place of systems consisting of a separator discharging a gas phase expanded in a turbo expander and a liquid phase let down in a single phase turbine or a valve. In this respect, turbines can be also seen as an efficient means of separation for oil and gas, if designed for this purpose.

Pressure let-down of a two-phase mixture through a valve is carried out by an isenthalpic transformation. It is characterised by the absence of energy transfer through the valve. However, pressure let-down of a two-phase mixture through a turbine approaches an isentropic process corresponding to an enthalpy reduction and therefore to energy removal from the fluid mixture. In addition to the production of energy, this type of pressure let-down also provides a lower outlet temperature and a higher outlet liquid flowrate than with a valve.

As the let-down process is not fully reversible, the enthalpy reduction is slightly smaller than that corresponding to a fully isentropic process. The ratio of these two enthalpy reductions defines the hydraulic efficiency of the turbine stage. This efficiency also determines the temperature reduction, the mass transfer and the energy provided by the stage, similarly to compression. This energy is not transmitted in full to the turbine shaft end due to the mechanical losses of some turbine components, such as bearings and seals. The ratio between the energy measured at the shaft end and the fluid energy defines the overall efficiency of a turbine. It is fairly obvious that turbine efficiency is lower than the hydraulic efficiency of the stage.

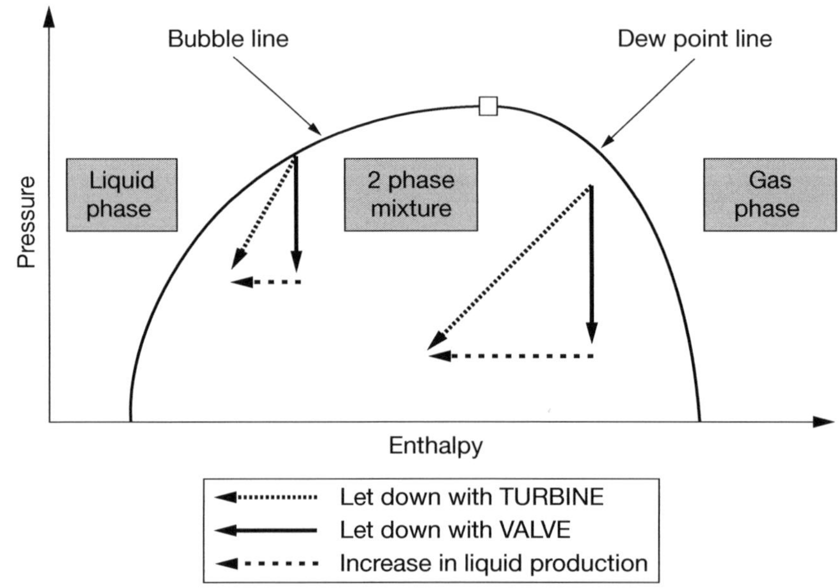

Figure 4.6

Two phase flow let down through valve and turbine

Several possible applications of multiphase turbines are presented below for oil and gas production as well as, downstream, for the petrochemical industry or any process facilities. There are many other cases where they could be used.

4.5.2.2 Upstream Applications

For upstream duties (oil and gas production), possible applications include the replacement of chokes valves by turbines, the use of turbo-pumps in fractured reservoirs with both high and low pressure wells to increase production pressure. Two-phase turbines can be used to reduce both the pressure and the temperature of high-pressure, high-temperature wells as well as, on a smaller scale, to ensure the production of electrical energy in isolated facilities (subsea or small platforms). Multiphase turbines can also be used in oil-gas separation units.

- **Replacement of a Choke Valve**

Choke valves are normally used to control the flow at well head. A two-phase (TP) turbine may be associated with a choke valve for controlling the flow and for recovering energy and additional hydrocarbon liquids. In that instance, the flow is controlled by the rotation speed of the turbine with further adjustment on the choke valve.

Charron *et al.* (2004) evaluated the benefits in the case of a condensate gas with a pressure let-down from 20 to 10 MPa abs. At a flowrate of 32 t/hr and a temperature of 40°C, the power recovery is about 300 kW with a 7% increase in liquid production. If the inlet temperature were to be 160°C, these results would approximately double.

The interests of a TP turbine are not negligible, especially with high-pressure and high-temperature (HP-HT) fields. Handling of high fluid pressure and temperature (for instance, above 100 MPa and 200°C) creates specific problems which can prevent the development of such fields. The pressure often has to be reduced below 10 MPa to allow gas and liquid treatment. With regard to the temperature, coolers have to be used to lower the temperature before treatment and export.

Unlike choke valves, the use of TP turbines allows power recovery as well as a significant reduction in the fluid temperature. For instance, with a condensate gas flow produced at well head at 80 MPa and 200°C with a production rate of 32 t/hr, the reduction of pressure to 7 MPa abs through a turbine provides some 1660 kW at the turbine shaft, *i.e.* 52 kW per t/hr. The temperature is reduced by 85°C as opposed to only 20°C in a valve. In the case of coolers designed for an outlet temperature of 100°C, using turbines enables the cooler duty to be reduced by more than 4 with a considerable impact on equipment weight and footprint.

In many HP-HT fields, recoverable power would be of the order of several tens of MW.

- **Boosting of Low Pressure Wells**

In fractured reservoirs, oil and gas are frequently produced at different pressure levels. As a result, the production pressure at the central platform is either lowered or the lowest pressure wells not produced. In the last case, a multiphase pump can be used to boost the production from the lowest pressure wells but external energy (generally electrical) has to be provided to the pump. Alternatively, because the highest pressure wells need throttling, it is preferable to use that lost energy to drive the pumps required at the lowest pressure wells. One simple arrangement consists in mounting two hydraulic sections in the same casing and on the same shaft, one producing energy (HP wells) and one absorbing it (LP wells) as shown in Figure 4.7.

The resulting intermediate production pressure (MP) is a function of the fluid temperature and the flow at each well as shown in Figure 4.8. According to this figure, the pressure of the LP well is increased by 1.2 and 1.7 MPa for HP well temperatures of 40 and 160°C respectively in the case of an equal production of LP and HP wells. The LP pressure rise is greater at a lower LP/HP flow ratio. The MP pressure is also a function of the fluid composi-

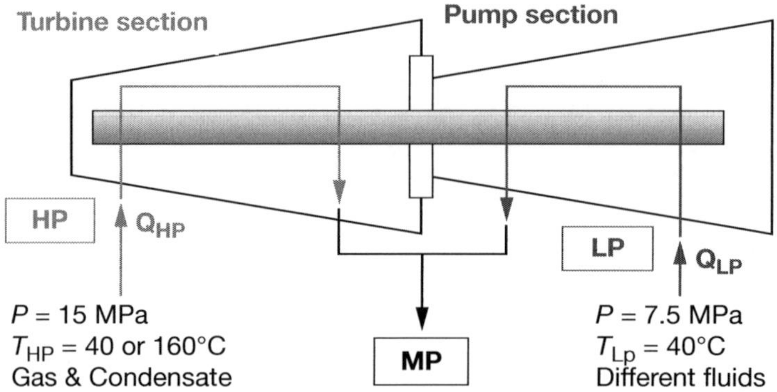

Figure 4.7

Two-phase pump and turbine for the production of fractured reservoirs

tion: for example, the LP pressure is raised from 1.2 to 1.8 MPa when the composition of the LP well is changed from a gas – condensate flow to an oil – associated gas flow (LP and HP flows equal and HP well at 40°C). The pressure rise is greater at a lower LP/HP flow ratio and at a higher HP well temperature.

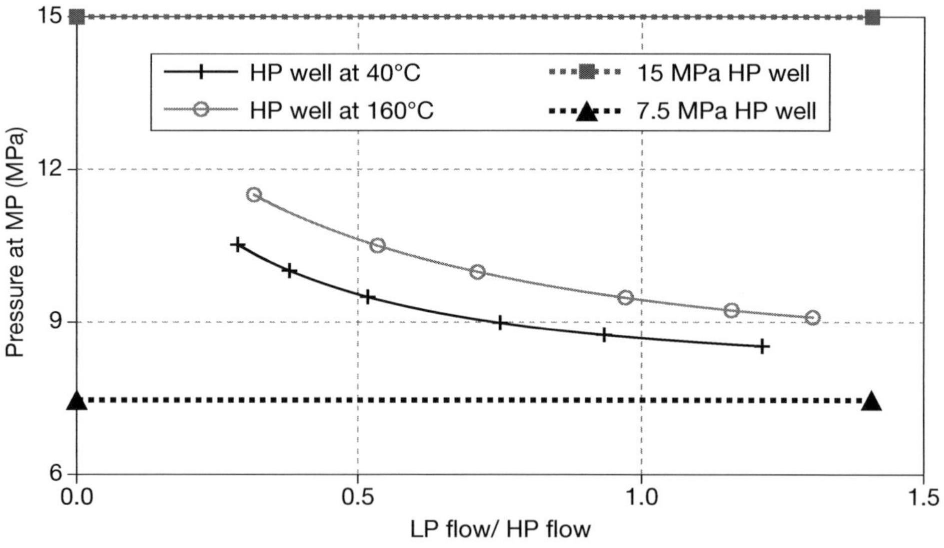

Figure 4.8

Two-phase pump and turbine for HP and LP wells
Variation of intermediate pressure, MP, versus flow ratio of LP and HP wells and temperature of HP well

- **Energy Production in Remote Areas**

Energy may be produced in remote areas from a two-phase flow when production pressure so allows. In this case, the two-phase flow is let down into a TP turbine driving an electrical generator with or without variable-speed depending on electrical requirements. As shown in the previous paragraphs, energy recovery varies mostly with the two-phase flowrate, the scale of the pressure let-down (pressure difference or pressure ratio), the gas fraction of the mixture (energy recovery increases with it), the gas composition (energy recovery increases with the reverse of the gas molecular weight) and the gas temperature (energy recovery increases with it).

These systems may be installed on remote platforms with limited energy supply or for fossil fuel savings (cost reduction on both fuel consumption and supply boats).

- **Hydrocarbon Liquid Separation**

Downstream of a well, the pressure of a crude oil mixture is reduced to remove most of the gas phase contained in the liquid phase. After separation, the gas is generally compressed for export or re-injection while the liquid phase is pumped into a pipe or to a storage facility. Separation is generally carried out in several stages to increase liquid production and to reduce the energy required by gas compression. The conventional process relies on density differences between the phases and gravity as driving forces. Separation is completed in large and heavy separators as relatively long residence times are needed to separate the fluids effectively.

TP turbines may be used for liquid stabilisation and separation. Separation of phases requires a special turbine design to ensure centrifugation. In a separation turbine, centrifugal forces are higher than in usual hydrocyclones. The liquid residence time is very short and this results in far more compact and lightweight equipment. Aboard floating production vessels, separation turbines are also less sensitive to motion. They also reduce the use of chemical additives to avoid foaming and production of an emulsion. In addition to these advantages, the drop in mixture enthalpy cools down the fluid and increases liquid recovery. These effects are highly dependent on the pressure and temperature conditions and the fluid composition.

Charron *et al.* (2004) have given calculation results illustrating the gains obtained with a TP turbine. They were made for a condensate gas with pressure letdown from 10 to 1 MPa abs, the liquid phase being stored at 0.2 MPa abs. With one separation stage (vessel at 1 MPa abs), using a TP turbine provides a 15% more liquid flow than with a valve and energy recovery of 700 kW. With two stages (vessels at 3 and 1 MPa abs), the liquid flow rises by 8% and the energy recovery is 350 kW.

Separation and stabilisation of crude oil by a turbine is advantageous as long as the possible pressure drop is large enough to provide the energy to drive the turbine and balance the various energy losses.

Three full-size two-phase separation turbines have been installed offshore on platforms in the Gulf of Mexico and a small three-phase prototype separation turbine was tested for 6 months in 2001-2002 in an onshore field in California, then on two platforms in the Gulf of Mexico [Rawlins & Ross, 2001, Oxley *et al.*, 2003].

4.5.2.3 Downstream Applications

Downstream applications in the petrochemical industry and any process facilities could include: replacement of two-phase let-down valves by turbines in refrigeration units to increase the cooling duty and reduce the compression duty (between condenser and evaporator), replacement of Joule-Thompson valves by turbines in Mixed Coolant Refrigerant circuits and in main gas liquefaction lines to increase the production of liquefied gas. The number of applications is considerable. Turbines are still used in a hydro-cracking process for energy recovery and in a gas processing unit for recovery of condensates. In general, any two-phase let-down through a turbine in a fluid process (petrochemical, chemical, geothermal) can provide an increase in cooling duty, energy recovery and liquid recovery.

- **Refrigeration Cycles**

A simple refrigeration cycle includes a compressor, a condenser, a pressure letdown valve and a chiller. Condensing temperature and pressure are determined by the cooling medium, while evaporating pressure and temperature are fixed by the compression ratio. Cycle efficiency may be improved by using more than one compression stage plus economisers. In most cases, the condensed refrigerant is let down from the bubble point thus entirely in two-phase flow.

TP turbines may be used in refrigeration cycles for energy recovery and to increase the cooling duty. In the case of refrigeration cycles with up to three stages of compression, comparisons have been made between cycles using letdown valves and TP turbines. In the case of a cycle with one compression stage, at a condensing temperature (CT) of 10°C, energy recovery is 5% of the compressor absorbed power and the increase in cooling duty is 7%. At a CT of 60°C, these figures become 12 and 25%. In the case of a cycle with three compression stages, at a CT of 60°C, these numbers become 6.5 and 8.7%. It is relatively clear from these examples that the benefit of a two-phase turbine increases with the condensing temperature and that it is greater with simplified cycles. TP turbines may be used to save energy, reduce the number of compression stages and therefore reduce space and weight.

- **LNG Production**

There are several ways to liquefy natural gas. The gas may be super-cooled at high pressure with the super-cooled liquid letdown to a pressure close to the storage condition or cooled near to the critical point. The cooling system may use one to three refrigeration cycles but also pure or mixed components. However, whatever the design of the LNG production plant, there is a requirement for two-phase letdown at some stage.

In case of a super-cooled liquid, the phase is commonly letdown with a JT valve. Installations are now gradually replacing JT valves with single-phase turbines, which in some cases leads to an increase in production of the order of 5%. Single-phase turbines are used until gas vaporisation occurs, above which point a two-phase turbine is required. If a two-phase turbine were to be used, an additional increase of 0.5 to 1% would be obtained.

- **Solvent Regeneration Cycles**

A solvent regeneration cycle is another typical example of use of a TP turbine. In this cycle, the gas is often absorbed in a solvent at high pressure and low temperature then the solvent is heated and depressurised to release the dissolved gas. The depressurisation process, usually carried out with a valve, could be performed by a two-phase turbine with a view to recovering energy and producing cold.

4.5.3 Technological Concepts

Several types of multiphase turbines are already available. However, these machines often operate in very specific conditions.

Turbo-expanders, extensively used in gas liquefaction and drying processes are only designed for two-phase flow forming downstream of the rotor inlet. They cannot accept a two-phase mixture at inlet owing to the extremely high peripheral velocity (about 400 m/s) and the risk of a rapid destruction of turbine internals by droplet erosion.

In this chapter, we only consider turbines which are designed to accommodate multiphase phase flows at inlet.

4.5.3.1 Impulse Type TP Turbines

Impulse type turbines appeared on the market some years ago. These machines are designed more to provide efficient separation of gas and liquid phases with a potential for energy recovery [Rawlins & Ross, 2001, Oxley *et al.*, 2003].

As indicated above, this type of turbine has already been tested on an onshore field in the US and three offshore fields.

4.5.3.2 Helico-Axial TP Turbine

A two-phase turbine of the helico-axial type has been designed by IFP and Union Pump and tested in IFP facilities. The features of this turbine are relatively similar to these of helico-axial pumps (Figure 4.9). These turbines are also multi-staged machines, with each stage including a rotor part and a stator part (the distributor). The turbine can accept any fraction of gas and liquid at inlet. It is relatively compact while offering good hydraulic performance, in particular when the liquid/gas density ratio is lower than 100 [Charron *et al.*, 2004].

At present this turbine is not designed to provide separation, but more to offer good energy recovery in two-phase flow.

4.5.4 Main Issues

The let-down of a two-phase flow comprises a number of major differences with the let-down of a single-phase flow.

In fluid flow let-down by a turbine, the reduction of fluid pressure is associated with mechanical work recovered on the machine shaft. Similarly to pressure boosting, no significant temperature change is created by reducing the reduction of an incompressible fluid, but a temperature reduction is observed with a compressible fluid. This reduction is smaller than the reduction corresponding to a fully insentropic process.

Figure 4.9

Helico-axial two-phase turbine prototype

The compressibility of the mixture and therefore the increase in volume flow is somewhere between those of the liquid and gas phases. This has to be taken into account in the design of a hydraulic cell to meet the pressure coefficient specification.

Heat transfer occurs between liquid and gas phases tending to attenuate the temperature variation of the gas phase. Its magnitude is determined by process conditions and the hydraulic efficiency of the turbine.

Mass transfer may occur between liquid and gas phases. The pressure let-down and mass transfer have a combined effect on the density ratio and the GVF (and therefore on two-phase efficiency) and on volume change (pressure coefficient specification). It should be noted that mass transfer may be positive or negative depending on the flow composition (gas condensation or liquid vaporisation).

Energy transfer from a fluid to the rotating shaft of a machine is not strictly the reverse of energy transfer from a rotor to a fluid considering the numerous differences in flow behaviour, such as blade incidence and diffusion losses, to mention but a few in single-phase flow. In addition, with two-phase mixtures, the gas bubble causing a performance drop reduction in a compression process will present totally different behaviour in a let-down process.

The velocity slip between liquid and gas in turn generates dissipation losses and therefore a reduction in the hydraulic efficiency of the turbine. The ratio of the two phase efficiency to the single-phase efficiency is called the two-phase efficiency multiplier or efficacy. This parameter, which is always smaller than 1, depends mainly on the gas-liquid density ratio and the GVF, both being defined on actual pressure and temperature conditions.

4.5.4.1 Diversity of Turbine Characteristics

A wide range of operating conditions are encountered in the applications described above. The gas fraction at inlet may vary from 0 (liquid at bubble point) to 1 (gas at dew point) with a large number of cases with a gas fraction of 0.9. The volume flow capacity may vary from approximately 100 m^3/hr to 20,000 m^3/hr.

The fluid pressure at inlet may vary from atmospheric pressure (LNG) to 100 MPa (High-pressure wells) with many cases around 10 MPa. The fluid temperature may vary from – 160°C for LNG to 250°C for high-temperature wells with a higher concentration of cases around 50°C.

For hydrocarbon production, the pressure let-down produces liquefaction of the gas phase varying from 0 to 25% of the gas flow depending on the production situation. Conversely, for refrigeration cycles, LNG production or solvent cycles, a fraction of the liquid phase is vaporised. This fraction is heavily dependent on the flow composition and pressure and temperature levels.

The combined effect of mass transfer and pressure let-down determines the volume change between the inlet and the outlet of a turbine section. In the case of gas liquefaction, the overall volume expansion is relatively small (generally, no more than 2 for hydrocarbon production). However, in the case of liquid vaporisation, the volume flow increase is relatively large (usually greater than 10 for refrigeration cycles).

The volume change through a hydraulic stage is determined by the volume expansion on a turbine section but also by the number of stages. It is often of the order of 1.1 in the case of gas liquefaction. However, it may rise to 2 in case of liquid vaporisation.

4.5.4.2 Energy Recovery and Production Schemes

Energy recovery from a well stream for separation, generation of electricity or other purposes has an obvious drawback. The energy is no longer available to export production through long-distance pipeline under natural pressure. Consequently, conventional production schemes, with separation and single-phase transportation are better-suited to multiphase turbines than deepwater long distance tie-backs.

References

Açikgöz M, França F, Lahey Jr RT (1992) An experimental study of three-phase flow regimes. Int. J. Multiphase Flow 18, 3, pp. 327-336.

Adejuyigbe B, Uvwo I, Ekpecham O, Scott SL, Lansangan R, Dutton R (2004) Investigation of three-phase flow measurement capabilities of a Coriolis flow meter. Presented at the 4th North American Conference on Multiphase Technology, Banff, Canada, Proceedings, B.H.R. Group pp. 161-176, June 2004.

Agar J, Farchy D (2002) Wet Gas Metering Using Dissimilar Flow Sensors: Theory and Field Trial Results. Paper SPE 77349 presented at the Annual Technical Conference & Exhibition, San Antonio, 29 Sept.-3 Oct. 2002.

Al Taweel A, Barlow S (1999) Field Testing of Multiphase Meters. Paper SPE 56583 presented at the Annual Technical Conference & Exhibition, Houston, 3-6 October 1999.

Alary V, Falcimaigne J (2002) Subsea water separation: a cost-effective solution for ultra deep water production, Paper 02PO presented at the 17th World Petroleum Congress, Rio, September 2002.

Alary V, Marchais F, Palermo T (2000) Subsea Water Separation and Injection: a Solution for Hydrates. Paper 12017 presented at the 2000 Offshore Technology Conference, Houston, May 2000.

Aly FA and Lee LL (1981) Self-Consistent Equations for Calculating the Ideal Gas Heat Capacity, Enthalpy, and Entropy. Fluid Phase Equilibria, 6, 169-179.

American Petroleum Institute (2002) Use of Subsea Wet-gas Flowmeters in Allocation Measurement Systems. API Recommended Practice 85, Dallas.

American Petroleum Institute (2004) State of the Art – Multiphase Metering. API publication 2566, Dallas.

American Petroleum Institute (2005) Recommended Practice for Measurement of Multiphase Flow. API RP 86, Dallas.

Andreussi P, Bendiksen K (1989) An investigation of void fraction in liquid slugs for horizontal and inclined gas-liquid pipe flow. Int. J. Multiphase Flow 15, 6, pp. 937-946.

Andreussi P, Sodini S, Faluomi V, Ciandri P, Ansiati A, Paone F, Battaia C, De Ghetto G (2003) Multiphase Ejector to Boost Production: First Application in the Gulf of Mexico. Paper OTC 15170 presented at the 2003 Offshore Technology Conference, Houston, 5-8 May 2003.

Andritsos N, Hanratty TJ (1987) Influence of interfacial waves in stratified gas-liquid flows. AIChE Journal 33, pp. 444-454.

Arnaudeau M (1982) Two phase pump developed by IFP. Ocean Industry, April 1982, pp 359-360.

Arnaudeau M (1988) Development of a Two-Phase Oil Pumping System for Evacuating Subsea Production Without Processing Over Long Distance. Paper OTC 5648 presented at the 20th Offshore Technology Conference, Houston, May 1988.

Atkinson D, Berard M, Segeral G (2000) Qualification of a Nonintrusive Multiphase Flow Meter in Viscous Flows, Paper SPE 63118, presented at the Annual Technical Conference & Exhibition, Dallas, 1-4 October 2000.

Austgen DM, Rochelle T, Peng X and Chen CC (1989) Model of Vapor-Liquid Equilibria for Aqueous Acid Gas – Alkanolamine Systems Using the Electrolyte-NRTL Equation. Ind. Eng. Chem. Res., **28**, 1060-1073.

Azzopardi B (1999) Phase separation at T-junction. Multiphase Science and Technology 11, pp. 223-329.

Baker O (1954) Simultaneous flow of oil and gas. Oil and Gas J. 53, pp. 185-195.

Baker LE, Pierce AC and Luks KD (1982) Gibbs Energy Analysis of Phase Equilibria. Soc. Pet. Eng. J., SPE **9806**, 731-742.

Ballard AL and Sloan ED (2002) The Next Generation of Hydrate Prediction: an Overview. Proceedings of the Fourth International Conference on Gas Hydrates, Yokohama, Japan, May 19-23 2002, 307-314.

Barnea D, Shoham O, Taitel Y (1980) Flow pattern characterization in two phase flow by electrical conductance probe. Int. J. Multiphase Flow 6, 5, pp. 387-397.

Barroux C, Duchet-Suchaux P, Samier P and Nabil R (2000) Linking reservoir and surface simulators: how to improve the coupled solutions. SPE paper **65159**, presented in SPE conference 24-25 October 2000 in Paris (France).

Bekkousha M, Khoory A, Metwalli M, Al Jawahiri H, Sakaria D, Al Rajabi S (2004) Multiphase Flow Meters – Field Trials and their Applications in Fields. Paper SPE 88742, presented at the 11th Int. Petroleum Exhibition and Conference, Abu Dhabi, 10-13 October 2004.

Benedict M, Webb GB and Rubin LC (1940) An Empirical Equation for Thermodynamic Properties of Light Hydrocarbons and their Mixtures. J. Chem. Phys., **8**, 334-345.

Bentolila Y, Constant M (2005) Physical approach of volumetric phase estimation based on mixture laws for multiphase flows, presented at the 12th Int. Conference on Multiphase Production Technology '05, Barcelone, proceedings, Vol 2, pp. 369-382.

Bergles AE, Collier JG, Delhaye JM, Hewitt GF, Mayinger F (1981) Two-Phase flow and heat transfer in the power and process industries. McGraw-Hill, Washington.

Bibet PJ (2005) Dunbar MPP: a good example of operation flexibility. Presented at the 7th Annual Multiphase Pump User Roundtable, Dept of Petroleum Engineering. TAMU, Houston, May 5, 2005.

Bird RB, Stewart WE, Lightfoot EN (1960) Transport phenomena. John Wiley & Sons, New York.

Boe A (1981) Severe slugging characteristics; Part I: Flow regime for severe slugging; part II: point model simulation study. Presented at Selected Topics in Two-Phase Flow, NTH, Trondheim, Norway, March 1981.

Bonizzi M, Issa R (2004) Numerical studies of three-phase gas/liquid/liquid slug flows. 5th International Conference on Multiphase Flow, ICMF'04. Yokohama, Japan, paper N° 523.

Brandt JU, Wietstock P (1992) Screw Pumps for Multiphase Pumping. Presented at the 4th EC Symposium, Berlin, 1992, proceedings. pp. 83-94.

Bratu C, Seince L (2005) New Progressing Cavity Pump (NPCP) for Multiphase and Viscous Liquid Production. Paper SPE 97833 presented at the 2005 SPE International Thermal Operations and Heavy Oil Symposium, Calgary, 1-3 November 2005.

Brauner N, Maron M (1992) Flow pattern transitions in two-phase liquid-liquid flow in horizontal tubes. Int. J. Multiphase Flow 18, 1, pp. 123-140.

Brauner N (2001) The prediction of dispersed flows boundaries in liquid-liquid and gas-liquid systems. Int. J. Multiphase Flow 27, pp. 885-910.

Brauner N, Ullman A (2002) Modeling of phase inversion phenomenon in two-phase pipe flows. Int. J. Multiphase Flow 28, pp. 1177-1204.

Brauner N, Ullman A (2004) Modelling of gas entrainment from Taylor bubbles. Part A: Slug flow. Int. J. Multiphase Flow 30, pp. 239-272.

Burkhardt J, Michie T (1979) Submerged Production System – A Final Report, Paper 3450 presented at the 11th Offshore Technology Conference, Houston, April-May 1979, proc. pp. 801-806.

Cai S, Toral H, Sinta D, Tajak M (2004) Experience in field tuning and operation of a multiphase meter based on neural net characterization of flow conditions. Presented at the Flomeko 2004 Conference, Proceedings, Beijing, 14-17 September 2004.

Campbell JM, Gas Conditioning and Processing. Volume 1: The Basic Principles. Ed. Norman: Campbell petroleum series, 1984.

Carroll JJ and Mather AE (1995) A Generalized Correlation for the Peng-Robinson Interaction Coefficients for Paraffin-Hydrogen Sulfide Binary Systems. Fluid Phase Equilibria, **105**, 221-228.

Cavett RH (1962) Physical Data for Distillation Calculations, Vapor-Liquid Equilibria. 27th Midyear meeting of the API Division of Refining, **42**, 3, 351-357.

Chapman WG, Gubbins KE, Jackson G and Radosz M (1990) New Reference Equation of State for Associating Liquids. Ind. Eng. Chem. Res., **29**, 1709-1721.

Charnock H (1955) Wind stress on a water surface. Quant. J. Roy. Met. Soc. 81, pp. 639-642.

Charron Y, Acloque E, Stilhé S (2003) Two Phase Helico-Axial Turbines. Paper SPE 84062 presented at the 2003 Annual Technical Conference and Exhibition, Denver, October 5-8, 2003.

Charron Y, Pagnier P, Marchetta E, Stilhé S (2004) Multiphase Flow Helico-Axial Turbine: Application and Performance. Paper SPE 88643 presented at the 11th Abu Dhabi International Petroleum Exhibition and Conference, Abu Dhabi, October 10-13, 2004.

Chisholm D (1977) Two-Phase Flow Through Sharp-Edged Orifices. Journal of Mechanical Engineering Science, 19, 3.

Cholet H (1997) Progressing Cavity Pumps. Editions Technip, Paris.

Cohen S L, Hanratty TJ (1968) Effect of waves at a gas-liquid interface on a turbulent air flow. J. Fluid Mech. 31, 3, pp. 467-479.

Colmenares J, Perry D, Guevara E, Padron A (1995) Multiphase Flow Metering of Heavy Crude – Oil Field Evaluation. Paper presented at the 7th Int. Conference on Multiphase Production, Proceedings, BHR Group, Cannes, 6-7 June 1995.

Conan C (2007) Analyse expérimentale et modélisation des écoulements liquide-liquide en conduite. Ph. D. dissertation, INPT, France.

Coniglio L, Trassy L and Rauzy E (2000) Estimation of Thermophysical Properties of Heavy Hydrocarbons through a Group Contribution Based Equation of State. Ind. Eng. Chem. Res., **39**, 5037-5048.

Cordner M, de Salis J (1995) The Nautilus Project: Development of a Subsea Multiphase Boosting System. Paper presented at the SUBSEA'96 Conference, December 12-13, 1996.

Coste P, Vilagines R (1993) Use of the Cathare 1D pump model to predict the two-phase characteristic of a pump for petroleum fluids. Presented at the European Two-Phase Flow Group Meeting, 7-10 June, Hanover, Germany, proceedings.

Cotton J., Stinessen K (1994) A Booster Station: Results of Extensive Testing and Post-Test Inspection. Paper OTC 7513 presented at the 26th Offshore Technology Conference, Houston, May 2-5, 1994.

Coutinho JAP, Andersen SI and Stenby EH (1995) Evaluation of Activity Coefficient Models in Prediction of Alkane Solid-Liquid Equilibria. Fluid Phase Equilibria, **103**, 23-29.

Crawford TJ, Weinberger CB, Weisman J (1985) Two phase flow patterns and void fractions in downward flow: steady-state flow patterns. Int. J. Multiphase Flow 11, 6, pp. 761-782.

Crouzier O (1978) Ecoulements diphasiques gaz-liquide dans les conduites faiblement inclinées par rapport à l'horizontale. Ph. D. dissertation. University of Paris VI, France.

Dahl E, Corneliussen S, Couput JP, Dykesteen E, Frøysa KE, Malde E, Moestue H, Moksnes PO, Scheers L, Tunheim H (2005) Handbook of Multiphase Metering. Norwegian Society for Oil and Gas Measurement Norwegian Society of Technical and Scientific Professionals, Olso.

Dal Porto DF, Larson LA (1996), Multiphase pumps trials demonstrate practical application for the technology. Paper SPE 36590 presented at the Annual technical meeting of the Society of Petroleum Engineers, Houston, 3-6 oct. 1996.

Darrigan PR, Skorping R, Bakken LE (1995) Multiphase Boosting – Field Experience and Future Developments. Paper presented at the Deep Offshore Technology Conference, 1995.

Davies RM, Taylor GI (1950). Proc. Roy. Soc. (London) 200, ser. A, pp. 375-390.

Decarre S, Fabre J (1997) Étude sur la prédiction de l'inversion de phase. Revue de l'Institut Français du Pétrole 52, 4, pp. 415-424.

de Crécy F (1983) Two-phase two pressure model – application to an axial pump model, (presented at the European Two-Phase Flow Group Meeting, 14-17 June, Zurich, Switzerland), proceedings, paper B14.

de Crécy F (1986) Modeling of stratified two-phase flow in pipes, pumps or other devices. Int. Journal of Multiphase Flow, Vol. 12, 3, pp. 307-323.

de Hemptinne JC, Ruffier-Meray V and Behar E (2006) Thermodynamic Modelling of Petroleum Fluids. Oil & Gas Science and Technology, accepted.

de Leeuw H (1997) Venturi meter performance in wet gas flow. Paper presented at the Int. Conference Multiphase, Cannes, France, June 18-20, 1997, proceedings pp. 567-582.

de Salis J, Heintze E, Charron Y (1999) Dynamic simulation of multiphase pumps. Multiphase'99, pp. 11-43.

de Salis J (2000) Pumps for Subsea Processing: Integration Issues. Paper 12013 presented at 2000 Offshore Technology Conference, Houston, May 2000.

Delaytermoz A (2000) Dunbar Phase II. Presented at the 2nd Annual Multiphase Pump User Roundtable, Dept of Petroleum Engineering, TAMU, Houston, May 4, 2000.

Derbyshire RTP, Hewitt GF, Nicholls B (1969) X-radiography of two-phase gas-liquid flow. AERE report M-1321.

do Vale OR, Garcia JE, Villa M (2002) VASPS Installation and Operation at Campos Basin. Paper 14003 presented at the 2002 Offshore Technology Conference, Houston, May 2002.

Dolan PJ, Goodridge R A, Leggate JS (1986) Development of a Twin Screw Pump for Multiphase Duties. Paper SPE 15881 presented at the European Petroleum Conference, London, 20-22 October 1986, proceedings pp. 293-298.

Donachie J (2004) Artificial Lift Application of Topside Multiphase Pump System. Offshore Engineer, Nov. 2004, pp. 49-51.

Duns Jr H and Ros NCJ (1963) Vertical flow of gas and liquid mixtures in wells. Proc. 6th World Petroleum Congress, Section II-Paper 22-PD6, Frankfurt, pp. 451-465.

Dykesteen E, Amdal J, Danielsen H, Flølo D, Grendstad J, Hide H, Moestue H, Torkildsen B (1995) Handbook of Multiphase Metering. Norwegian Society for Oil and Gas Measurement, Norwegian Society of Chartered Engineers, Olso.

Edmister, WC (1958) Applied Hydrocarbon Thermodynamics, Part 4: Compressibility factors and Equations of State. Pet. Refiner, **37**, 4, 173-179.

Engelman E, Torp TA (1990) The Poseidon Project and the Future of Multiphase Production. Paper presented at the *Offshore Northern Seas Conference*, Stavanger, proceedings.

Esson AL, Cohen DJ (1996) The development of a gas-handling downhole pump. IMechE C508/026/96.

Falcimaigne J, Durando P, Loupias M, Vilagines R (1994) Multiphase Rotodynamic Pumps Extend their Operating Capabilities. Paper SPE 28882 presented at the European Petroleum Conference, London, October 1994.

Falcone G, Hewitt G, Alimonti C, Harrison B (2002) Multiphase Flow Metering: Current Trends and Future Developments. Paper SPE 74689, Distinguished Author Series, Journal of Petroleum Technology, April 2002.

Ferschneider G (1983) Ecoulement diphasique gaz-liquide à poches et à bouchons en conduites. Revue Inst. Fr. Pétrole 38, pp. 153.

Firoozabadi A (1999) Thermodynamics of Hydrocarbon Reservoirs, McGraw-Hill.

Flores JG (1997) Oil-water flow in vertical and deviated wells. Ph. D. dissertation. University of Tulsa, USA.

Fredenslund A, Gmehling J and Rasmussen P (1977) Vapour-Liquid Equilibria using UNIFAC. Elsevier, Amsterdam.

Fredenslund A, Jones RL and Prausnitz JM (1975) Group Contribution Estimation of Activity Coefficients in Nonideal Liquid Mixtures. AIChE. J., **27**, 1089-1099.

Furuya O (1985) An analytical model for prediction of two-phase non condensable flow pump performance. Journal of Fluid Engineering, Vol 107, pp. 139-147.

Gardel A, Rechsteiner GF (1970) Les pertes de charge dans les branchements en té des conduites de section circulaire. Bulletin Technique de la Suisse Romande 25, pp. 363-391.

Garrett M, Drinkard D (2002) Ceiba Field Development: Fast Track to Early Cash Flow Management's Perspective. Paper OTC 14080, presented at the 2002 Offshore Technology Conference, Houston, May 2002.

Gas Processing Suppliers Association (1998) Engineering Data Book, 11th edition. GPSA, Tulsa, OK.

Gié P, Buvat P, Bratu C, Durando P (1992) Poseidon Multiphase Pump: Field Tests Results. Paper OTC 7037 presented at the 24th Offshore Technology Conference, Houston, May 1992.

Golan LP, Stenning AH (1969) Two-phase vertical flow maps. Proc. Inst. Mech. Eng. & Inst. Chem. Eng. Fluid Mech. & Meas. in two-phase flow syst. Symp. 184, PT 3C, pp. 108-114.

Goodfellow R, Webb A, Harbonn J (1974) Subsea Experience Gained at Zakum. Paper OTC 1942 presented at the 6th Offshore Technology Conference, Houston, May 6-8, 1974.

Govier GW and Aziz K (1972) The flow of complex mixtures in pipes. Van Nostrand Reinhold Co, New York.

Grayson HG and Streed CW (1963) Vapor-Liquid Equilibria for High Temperature, High Pressure Hydrogen-Hydrocarbon Systems. Proceedings of the 6th World Petroleum Congress, Frankfurt/ Main III, Paper 20-PD7, 233-245.

Gregory GA, Nicholson MK, Aziz K (1978) Correlation of the liquid volume fraction in the slug for horizontal gas-liquid slug flow. Int. J. Multiphase Flow 4, pp. 33-39.

Griffits P, Wallis GB (1961) Two-phase slug flow. J. Heat Trans. 83, pp. 307-320.

Guet S, Decarre S, Henriot V, Liné A (2006) Void fraction in vertical gas-liquid slug flow: influence of liquid slug content. Chem. Eng. Sci. Vol 61, 22, pp. 7336-7350.

Hajiloo M, Chang BH et Mills AF (2001) Interfacial shear in downward two-phase annular co-current flow. Int. J. Multiphase Flow 27, pp. 1095-1108.

Hale CP (2000) Slug formation, growth and decay in gas-liquid flows. Ph. D. dissertation. Imperial College of Science, Technology & Medicine, London, England.

Hall A, Reader-Harris M, Millington B (2000) A study of the performance of Venturi meters in multiphase flow. Presented at the 2nd North American Conference on Multiphase Technology, Banff, Canada, 21-23 June 2000, proceedings pp. 237-265.

Harvey AH and Levelt Sengers JMH (1990) Correlation of Aqueous Henry's Constants from 0°C to the Critical Point. AIChE J., **36**, 4, 539-546.

Hasebe B, Hall A, Smith B, Brady J, Mehdizadeh P (2004) Field Qualification of Four Multiphase Flowmeters on North Slope, Alaska. Paper SPE 90037 presented at the Annual Conference and Exhibition, Houston, 26-29 September 2004.

Hayden JG and O'Conell PO (1975) Generalized Method for Predicting Second Virial Coefficients. Ind. Eng. Chem. Proc. Des. Dev., 14, 3, 209-216.

Hayes J, Carroll W, Fothergill D, Prendergast G (1985) Hydrocyclones for Treating Oily Water: Development and Field Testing in Bass Strait. (presented at the OTC, Houston) paper 5079, proceedings, Houston, 6-9 May 1985.

Heintzé E, de Salis J, Charron Y (1999) Dynamic simulation of multiphase flow pumps. Presented at the Int. Conference Multiphase 99, Cannes, France, proceedings pp.11-43.

Henriot V, Courbot A, Heintzé E and Moyeux L (1999) Simulation of Process to Control Severe Slugging: Application to Dunbar Pipeline. SPE paper **56461**, presented in SPE conference 3-6 October 1999 in Houston (Texas).

Hewitt G F, Roberts D N (1969) Studies of two-phase flow patterns by simultaneous x-rays and flash photography. AERE report M-2159.

Hill TJ (1990) Gas injection at riser bases solves slugging flow problems. OGJ, February 26, 1990, pp. 88-92.

Hinze JO (1955) Fundamentals of the hydrodynamic mechanism of splitting in dispersion process. AIChE Journal 1,3, pp. 289-295.

Homer CJ (1985) Theoretical studies of pump performance in two-phase flow. Paper E4 presented at the European Two-Phase Flow Group Meeting, Marchwood, June 4-7,1985.

Horn T, Bakke W Eriksen G (2003) Experience in Operating World's first Subsea Separation and Water Injection Station at Troll Oil Field in the North Sea. Paper 15172 presented at the 2003 Offshore Technology Conference, Houston, May 5-8 2003.

Hout R, Barnea D, Shemer L (2002) Transitional velocities of elongated bubbles in continuous slug front. Int. J. Multiphase Flow 28, pp. 1333-1350.

Hubbard MG and Dukler AE (1966) The characterization of flow regimes for horizontal two-phase flow. Proc. Heat Transfert and Fluid Mechanics Institute, Stanford university Press CA.

Huron MJ and Vidal J (1979) New Mixing Rules in Simple Equations of State for Representing Vapour-Liquid Equilibria of Strongly Non-ideal Mixtures. Fluid Phase Equilibria, 3, 255-271.

International Commission on Radiological Protection (1991) Publication 60: 1990 Recommendations of the International Commission on Radiological Protection. Elsevier, Annals of the ICRP, Vol 21/ 1-3. Several other documents are also of interest. See the list of publications on the Internet www.icrp.org

International Standards Organisation (2002) Allocation of Gas and Condensate in the Upstream Area (2002), Draft version of report by TC 193 – SC3 – WG1.

Ishii M (1975) Thermo-fluid dynamic theory of two-phase flow. Eyrolles, Paris.

Jansen FE (1990) Elimination of severe slugging in a pipeline-riser system. Ph. D. Dissertation, University of Tulsa, USA.

Johansen GA, Jackson P (2004) Radioisotope Gauges for Industrial Process Measurements, Wiley, Chichester.

Jones Jr OC, Zuber N (1975) Interrelation between void fraction fluctuations and flow patterns in two-phase flow. Int. J. Multiphase Flow 2, 3, pp. 273-306.

Kay WB (1936) Density of Hydrocarbon Gases and Vapors at High Temperatures and Pressures. Ind. Eng. Chem, **28**, 1014-1019.

Kesler MG and Lee BI (1976) Improve Prediction of Enthalpy of Fractions. Hydrocarbon Proc., **53**, 3, 153-158.

Khoory A, Bekkousha M, Pinguet B, Turna E, Koshy T (2003) Multiphase Flowmeter and Production Logs Diagnose Well Response in an Onshore Field, Abu Dhabi. Paper SPE 81534, presented at the 13th Middle East Oil Show & Conference, Bahrain, 9-12 June 2003.

Kontogeorgis GM, Voutsas E, Yakoumis I and Tassios DP (1996) An Equation of State for Associating Fluids. Ind. Eng. Chem. Res., **35**, 4310-4318.

Kordas A, Magoulas K, Stamakis S and Tassios D (1995) Methane-Hydrocarbon Interaction Parameters Correlation for the Peng-Robinson and the t-mPR Equation of State. Fluid Phase Equilibria, **112**, 33-44.

Korolev S, de Salis J, Birnov M (1999) Oil Field Development in Western Siberia: A Multiphase Pump Case Study, Paper presented at the Int. Multiphase'99 Conference, Cannes, June 1999, proceedings pp. 45-54.

Kowalski JE (1987) Wall and interfacial shear stress in stratified flow in a horizontal pipe. AIChE Journal 33, pp. 274-281.

Lee BI and Kesler MG (1975) A Generalized Thermodynamic Correlation Based on Three-Parameter Corresponding States. AIChE. J., **21,** 3, 510-527.

Legget R, Boling D, Powers B, Shehata K, Halvorsen M, Aboelenain A (1996) Multiphase Flowmeter Successfully Measures Three-Phase Flow at Extremely High Gas-Volum Fractions – Gulf of Suez, Egypt. Paper SPE 36837 presente at the European Petroleum Conference, Milan, 22-24 October 1996.

Leibovici CF (1993) A Consistent Procedure for the Estimation of Properties Associated to Lumped Systems, Fluid Phase Equilibria, **84**, 1-8.

Leibovici CF, Barker JW and Waché D (1998) A Method for Delumping the Results of a Compositional Reservoir Simulation. SPE paper **49068**, presented in SPE conference 27-30 September 1998 in New-Orleans (USA, Louisiana).

Leporcher E, Taiani S (1995) Multiphase Pumping: the Lessons of Long-Term Field Testing. Paper SPE 30661 presented at the SPE Annual Technical Conference, Dallas, October 1995.

Leporcher E, Lezeau P, Johansen O (1999) Which mudline technology to deploy for multiphase transportation in deep offshore operations? Paper presented at Deep Offshore Technology conference, Stavanger, Oct. 1999.

Leporcher E, Delaytermoz A, Renault JF, Gerbier A, Burger O (2001) Deployment of Multiphase Pumps on a North Sea Field, Paper SPE 71536 presented at the SPE Annual Technical Conference and Exhibition, New Orleans, 30 Sept.-3 Oct. 2001.

Li ZY, Kitami H, Kawaoto H, Watanabe T (2004) A turbine flow meter for multiphase flow. Paper SPE 88741 presented at the 11th Abu Dhabi Int. Petroleum Conference, 10-13 October 2004, Abu Dhabi.

Lin ZH (1980) Two-phase flow measurements with orifices. In: Encyclopedia of Fluid Mechanics (Cheremisinoff NP, ed). Elsevier Science & Technology Books, Vol 3, pp. 841-862.

Lin PY, Hanratty TJ (1987) Detection of slug flow from pressure measurements. Int. J. Multiphase Flow 13, 1, pp. 13-21.

Liu YD and Watanasiri S (1999) Succesfully Simulate Electrolyte Systems. Chem. Eng. Prog., 95, 10, 25-42.

Lohrentz J, Bray BG and Clark CR (1964) Calculation of Viscosities of Reservoir Fluids from their Compositions. Journal of Petroleum Technology, 1171.

Lovick J, Angeli P (2004) Experimental studies on the dual continuous flow pattern in oil-water flows. Int. J. Multiphase Flow 30, pp. 139-157.

Lynch J (1991) Techniques de débitmétrie polyphasique non intrusive – Revue bibliographique. Revue de l'Institut Français du Pétrole, Vol 46, n° 1, jan-fev 1991.

Macleod DB (1923) On a Relation Between Surface Tension and Density. Trans. Faraday Soc., **19**, 38-43.

Malinowski MS (1975) An experimental study of oil-water and air-oil-water flowing mixtures in horizontal pipes. Ph. D. Dissertation, University of Tulsa, USA.

Malnes D (1982) Slug flow in vertical, horizontal and inclined pipes. Report IFE/KR/E-83/002, V. Inst. For Energy Technology, Kjeller, Norway.

Manolis IG (1995) High pressure gas-liquid slug flow. Ph. D. dissertation. Imperial College of Science. Technology & Medicine, London, England.

Marrelli JD, Hatton GJ, Siddiqui F, Pepin LL, Helm DA (1992), Continuous Determination of Oil Pipeline Watercut, Salinity and API Gravity, Regardless of Gas Fraction: the Microwave Watercut Monitor. Paper SPE 22401 presented at the International Petroleum Engineering Meeting, Beijing, 24-27 March 1992, proceedings, pp. 739-752.

Martin CS (1973) Transition from bubbly to slug flow of a vertical downward air-water flow. Proc. ASME Symposium, Atlanta.

Mathias PM and Copeman TW (1983) Extension of the Peng-Robinson Equation of State to Complex Mixtures: Evaluation of the Various Forms of the Local Composition Concept. Fluid Phase Equilibria, **13**, 91-108.

Mehdizadeh P, Farchy D (1995) Multiphase Flow Metering Using Dissimilar Flow Sensors – Theory and Field Trials Results. Paper SPE 29847 presented at the Middle East Oil Show, Bahrain, 11-14 March 1995.

Michelsen ML (1982) The Isothermal Flash Problem. Part I: Stability. Fluid Phase Equilibria, **9**, 1-19.

Michelsen ML (1990) A Modified Huron-Vidal Mixing Rule for Cubic Equations of State. Fluid Phase Equilibria, **60**, 213-219.

Mirza KZ, Wild AG (1997) Key Advantages of the Progressing Cavity Pump in Multiphase Transfer Applications. Paper SPE 38812 presented at the 1997 SPE Annual Technical Conference, San Antonio, 5-8 October 1997, proceedings, pp. 623-630.

Mohamed P, Al-Saif K (1998) Field Trial of a Multiphase Flow Meter. Paper SPE 4916 presented at the Annual Conference and Exhibition, New Orleans, 27-30 September 1998.

Montel F and Gouel PL (1984) A new lumping scheme of analytical data for compositional studies. SPE paper **13119**, presented in the 59th Annual Technical Conference and Exhibition, 16-19 September 1984, Houston, Texas (USA).

Mouly (1979) Ecoulements stratifiés de gaz naturel et d'huile en conduite pétrolière. Ph. D. dissertation, University of Paris VI, France.

Moysan JM, Paradowski H and Vidal J (1986) Prediction of Phase Behaviour of Gas Containing Systems with Cubic Equations of State. Chem. Eng. Sci., **41**, 8, 2069-2074.

Müller-Link D, Brandt JU, Reichwage M, Schröder G (2002) Wet gas compression with twin-screw multiphase pumps. Paper presented at the European Congress on Fluid Machinery for the Oil, Gas and Petrochemical Industry, The Hague, October 31, 2002, proceeding published by the Institution of Mechanical Engineers, London.

Müller-Link D, Rohlfing G, Reichwage M, (2004) Subsea multiphase boosting – a full size test rig. Paper presented at the 4th North American Conference on Multiphase Technology, Banff, Canada, June 3-4, 2004, pp. 347-354.

Murdock JW (1962) Two-phase flow measurement with orifices. Journal of Basic Engineering, Dec 1962.

Mus E, Toskey E, Bascoul S, Barber E (2002a) Development Well Testing Enhancement Using a Multiphase Flowmeter. Paper SPE 77769, presented at the Annual Technical Conference & Exhibition, San Antonio, 29 sept.-2 October 2002.

Mus E, Toskey E, Bascoul S, Norris R (2002b) Added Value of Multiphase Flowmeter in Exploration Well Testing. Paper SPE 80283, Production & Facilities Journal, November 2002.

NEL (1999) The Relative Permittivity of Oils. National Engineering Laboratory, Glasgow, report 251/99, 1st June 1999.

Ngai C, Brown M, Mehdizadeh P (1997) Performance Test of a High Gas Volume Fraction Multiphase Meter in a Producing Field. Paper SPE 38784, presented at the Annual Technical Conference & Exhibition, San Antonio, 5-8 October 1997.

Nicklin DJ, Wilkes JO, Davidson JF (1962) Two phase flow in vertical tubes. Trans. Instn. Chem. Engrs. 40, pp. 61-68.

Nicolaisen H, Rasmussen P and Sorensen JM (1993) Correlation and Prediction of Mineral Solubilities in the Reciprocal Salt System (Na^+, K^+)(Cl^-, $SO_4^=$) – H_2O at 0-100°C. Chem. Eng. Sci., **48,** 18, 3149-3158.

Nuland S, Skarsvåg K, Sæther G, Fuchs P (1991) Phase fractions in three-phase gas-oil-water flow. Presented at the 5th International Conference in Multiphase Production, Cannes, Proceedings, Elsevier Science Publishers, 1999, pp. 3-30.

Nuland S, Vilagines R (2001) Gas hydrate slurry flow – A flow modeller looks at the state of slurry rheology modelling. Paper presented at the BHRG Multiphase International Conference, Cannes, June 2001.

Nydal OJ, Andreussi P (1991) Gas entrainment in a long liquid slug advancing in a near horizontal pipe. Int. J. Multiphase Flow 17, 2, pp. 179-189.

Nydal OJ, Banerjee S (1996) Dynamic slug tracking simulations for gas-liquid flow in pipelines. Chem. Eng. Comm, 141-142 m, pp. 13-39.

Odozi UA (2000) Three-phase gas/liquid/liquid slug flow. Ph. D. dissertation, Imperial College of Science, Technology & Medicine, London, England.

Oshinowo T, Charles ME (1974) Vertical two-phase flow in vertical tubes. Trans. Inst. Chem. Eng. 52, pp. 25-35.

Oxley K, Bennett J, Fremin L, Taylor J, Ross G (2003) RST's Mission to Mars – The First Commercial Application of Rotary Separator Turbine Technology. Paper OTC 15357 presented at the Offshore Technology Conference, Houston, May 5-8, 2003.

Pal R (2001) Novel viscosity equations for emultions of two immiscible liquids. J. Rheol. 45, 2, pp. 509-519.

Pan L (1996) High pressure three-phase (gas-liquid-liquid) flow. Ph. D. dissertation, Imperial College of Science, Technology & Medicine, London, England.

Passut CA and Danner RP (1972) Correlation of Ideal Gas Enthalpy, Heat Capacity and Entropy. Ind. Eng. Chem. Process Des. Dev., **11**, 4, 543-546.

Pedersen KS, Thomassen P and Fredenslund A (1984) Thermodynamics of Petroleum Mixtures Containing Heavy Hydrocarbons - 1. Phase Envelope Calculations by Use of the Soave-Redlich-Kwong Equation of State. Ind. Eng. Chem. Process Des. Dev., **23**, 163-170.

Peebles FN, Garber HJ (1953). Chem. Eng. Progr. 49, pp. 88-97.

Peneloux A, Rauzy E and Freze R (1982) A Consistent Correction for Redlich-Kwong-Soave Volumes. Fluid Phase Equilibria, **8**, 7-23.

Peng D Y and Robinson D B (1976) A New Two-Constant Equation of State. Ind. Eng. Chem. Fundam., **15**, 59-64.

Pershukov VA, de Salis J, Cordner M (2001) Multiphase Pumps for 250 000 bbl/d Field Development. Paper SPE 71562 presented at the SPE Annual Technical Conference, New Orleans, September 30-October 3, 2001.

Pessoa R, Prado M (2001) Two-Phase Flow Performance for Electrical Submersible Pump Stages. Paper SPE71552 presented at the 2001 SPE Annual Conference and Exhibition, New Orleans, 30 September-3 October 2001.

Petroleum Engineering Handbook, Society of Petroleum Engineers, chapter 22, 1998.

Peysson Y, Maurel P, Vilagines R (2003) Hydrate transportability in multiphase flow. Paper presented at the BHRG Multiphase International Conference, San Remo, June 2003.

Peysson Y, Nuland S, Maurel P, Vilagines R (2003) Flow of Hydrates Dispersed in Production Lines. Paper SPE 84044 presented at the 2003 Annual Technical Conference and Exhibition, Denver, October 5-8, 2003.

Pitzer KS (1973) Thermodynamics of Electrolytes I: Theoretical Basis and General Equations. J. Phys. Chem., **77**, 2, 268-277.

Pitzer KS (1995) Thermodynamics, 3rd ed., McGraw-Hill, New-York.

Plocker U, Knapp H and Prausnitz J (1978) Calculation of High-Pressure Vapor-Liquid Equilibria from a Corresponding-States Correlation with Emphasis on Asymmetric Mixtures. Ind. Eng. Chem. Process Des. Dev., **17**, 3, 324-332.

Poling BE, Prausnitz JM and O'Connell JP (2001) The Properties of Gases and Liquids. 5th ed., McGraw-Hill, 2001.

Prausnitz JM, Lichtenthaler RN and Gomes de Azevedo E (1999) Molecular Thermodynamics of Fluid Phase Equilibria. 3rd ed., Prentice Hall Int.

Rachford HH Jr. and Rice JD (1952) Procedure for Use of Electronic Digital Computers in Calculating Flash Vaporization Hydrocarbon Equilibrium. Trans. Soc. Pet. Eng., **195**, 327.

Rackett HG (1970) Equation of State for Saturated Liquids. J. Chem. Eng. Data, **15**, 4, 514-517.

Ramberg RM, Bakken LE (1997) Multiphase pump operability in petroleum applications. Presented at the Int. Conference Multiphase, Cannes, France, June 18-20, 1997, proceedings pp. 199-212.

Rawlins C, Ross G (2001) Field Test Results of a Rotary Separator Turbine on the Ram/Powell TLP. Paper OTC 13218 presented at the Offshore Technology Conference, Houston, April 30-May 3, 2001.

Renon H and Prausnitz JM (1968) Local Compositions in Thermodynamic Excess Functions for Liquid Mixtures. AIChE. J., 14, 1, 135-144.

Riazi MR (2005) Characterization and Properties of Petroleum Fractions. 1st ed., ASTM International.

Riazi MR and Daubert TE (1980) Simplify Property Predictions. Hydrocarbon Processing, 59, 3, 115-116.

Riazi MR and Daubert TE (1987) Characterization Parameters for Petroleum Fractions. Ind. Eng. Chem. Res., 26, 755-759.

Riazi MR, Al-Sahhaf TA and Al-Shammari MA (1998) A Generalized Method for Estimation of Critical Constants. Fluid Phase Equilibria, 147, 1-6.

Riiser K, Fabre J, Suzanne C (1992) Gas entrainment at the rear of Taylor-bubble. 29th meeting of European Two-Phase Flow Group, Stockholm, 1-3 June 1992.

Sarsha MM, Beg NA (2000) The application and performance of jet pumps in oil and gas production handling multiphase flow. Presented at the 2nd North American Conference on Multiphase Technology, Banff, Canada, proceedings. pp. 369-385.

Scheers AM (1998) Multiple energy gamma ray absorption (MEGRA) techniques applied to Multiphase Flow Metering. Paper presented at the IBC Conference "The Future of Multiphase Metering", March 1998, London, Proceedings, IBC, London.

Schmidt H, Loth R (1994) Predictive methods for two-phase flow pressure loss in tee junctions with combining conduits. Int. J. Multiphase Flow, 20, 4, pp. 703-720.

Shen J, Riley R (1997) Field Evaluation of a Multiphase Meter in Well-Testing Operation. Paper SPE 37436 presented at the Production Operations Symposium, Oklahoma City, 9-11 March 1997.

Sinquin A, Palermo T and Peysson Y (2004) Rheological and Flow Properties of Gas Hydrate Suspensions. Oil & Gas Science and Technology – Rev. IFP, Vol. 59, No. 1, pp. 41-57.

Sekogushi K, Mori K, Kaji M, Nakazatomi M, Shimizu H (1996) Interfacial profiles and flow characteristics in vertical downward two-phase plug and foam flows. Chem. Engng Comm 141-142, pp. 415-441.

Serica C (1990) Two-phase flow in low velocity hilly terrain pipelines. Ph. D. Dissertation, University of Tulsa, USA.

Setzmann U and Wagner W (1991) A New Equation of State and Tables of Thermodynamic Properties for Methane Covering the Range from the Melting Line to 625 K at Pressures up to 100 MPa, J. Phys. Chem. Ref. Data, **20**, 6, 1061-1155.

Sevik M, Park SH (1965) The splitting of drops and bubbles by turbulent fluid flow. J. Fluids Eng. 95, pp. 53-60.

Shean AR (1976) Pressure drop and phase fraction in oil-water-air vertical pipe flow. Master Dissertation, Massachusetts Institute of Technology, USA.

Shoham O (1983) Flow pattern transition and characterization in gas-liquid two phase flow in inclined pipes. Ph. D dissertation, Tel Aviv University, Israel.

Sloan ED (1998) Clathrate hydrates of natural gases. 2nd ed., Marcel Dekker, New-York.

Smith T (1995) The Application of Multiphase Metering on Oil and Gas Gathering Platforms. Paper SPE 30662 presented at the Annual Technical Conference & Exhibition, Dallas, 22-25 October 1995.

Soave G (1972) Equilibrium Constants for a Modified Redlich-Kwong Equation of State. Chem. Eng. Sci, **27**, 1197-1203.

Songhurst BW, Edwards W (1989) Subsea separation: an economic method for submarginal fields. Paper 5922 presented at the 21st Offshore Technology Conference, Houston, May 1-4 1989, proc. Vol 1, pp. 405-416.

Soreide I and Whitson CH (1992) Peng-Robinson Predictions for Hydrocarbons, CO_2, N_2, and H_2S with Pure Water and NaCl Brine. Fluid Phase Equilibria, **77**, 217-240.

Spedding PL and Nguyen VT (1980) Regime maps for air water two phase flow. Chem. Eng. Sci. 35, 4, pp. 779-793.

St-Arnaud JM, Okambawa R, Bose TK, Le Noë O (1997) Comparison between a pressure-volume-temperature method (Z meter) and the dielectric constant method: application to the compressibility factor of five gaseous mixtures. High Temperatures – High Pressures, vol. 29, 1997, pp. 207-212.

Steven RN (2001) Wet gas metering with a horizontally mounted Venturi meter. Flow Measurement and Instrumentation 12,5-6, pp. 361-372.

Taitel Y (1986) Stability of severe slugging. Int. J. Multiphase Flow 12, 2, pp. 203-217.

Taitel Y and Duckler AE (1976) A model for predicting flow regime transitions in horizontal and near horizontal gas-liquid flow. AIChE Journal 22, 1, pp. 152-160.

Taitel Y, Barnea D, Duckler AE (1980) Modelling flow pattern transitions for steady upward gas-liquid flow in vertical tubes. AIChE Journal 26, 3, pp. 345-354.

Taitel Y, Vierkandt S, Shoham O, Brill JP (1990) Severe slugging in a pipeline-riser system, experiments and modelling. Int. J. Multiphase Flow 16, 1, pp. 57-68.

Taitel Y, Barnea D, Brill JP (1995) Stratified three phase flow in pipes. Int. J. Multiphase Flow 21, 1, pp. 53-60.

Tasselli S., Baldassarre L., Pucci P. (2003) Subsea compression technology. Presented at the Offshore Mediterranean Conference and Exhibition, Ravenna, Italy, March 26-28, 2003.

Tollefsen J, Hammer EA (1998) Capacitance sensor design for reducing errors in phase concentration measurements. Flow Measurement and Instrumentation 9, 1, pp. 25-2.

Toma P, Ridley R, Renenberg T, Chen M (2000), Development and testing of a novel gas-liquid flowmeter combining pressure fluctuations generated by vertical and horizontal flow patterns. Presented at the 2nd North American Conference on Multiphase Technology, Banff, Canada, 21-23 June 2000, proceedings. pp. 327-352.

Toral H, Cai S, Akartuna E, Stothard, Jamieson A. (1998) Field tests of the Esmer multiphase flowmeter. Paper presented at the North Sea Flow Measurement Workshop, 26-29 October 1998, Perthshire, Scotland.

Trallero JL (1995) Oil-water flow patterns in horizontal pipes. Ph. D. Dissertation, University of Tulsa, USA.

Tsonopoulos C (1974) Empirical Correlation of Second Virial Coefficients. AIChE. J., **20**, 263-272.

Tutu NK (1982) Pressure fluctuations and flow pattern recognition in vertical two-phase gas-liquid flows. Int. J. Multiphase Flow 8, 4, pp. 443-447.

Tuss B, Perry D, Shoup G (1996) Field Tests of the High Gas Volume Fraction Multiphase Meter. Paper SPE 36594, presented at the Annual Technical Conference & Exhibition, Denver, 6-9 October 1996.

Twu CH (1984) An Internally Consistent Correlation for Predicting the Critical Properties and Molecular Weights of Petroleum and Coal-tar Liquids. Fluid Phase Equilibria, **16**, 137-150.

Twu CH, Coon JE and Cunningham JR (1995) A New Generalized Alfa Function for a Cubic Equation of State. Part I. Peng-Robinson equation, Fluid Phase Equilibria, **105**, 49-59.

Twu CH, Tassone V and Sim WD (2004) Accurately Predict the VL of Thiol-Hydrocarbon Mixtures. Chem. Eng. Progress, **100,** 9, 39-45.

UK Department of Trade and Industry (2003) Guideline Notes for Petroleum Measurements under the Petroleum (Production) Regulations, Oil and Gas Division, Issue 7, Dec. 2003, London.

Ulaby F, Moore R, Fung A (1986) Microwave Remote Sensing: Active and Passive (in three volumes), Vol III, app. E, Artech House, Dedham, MA, USA.

Ungerer P and Batut C (1997) Prediction of the Volumetric Properties of Hydrocarbons with an Improved Volume Translation Method. Revue de l'Institut Français du Pétrole, **52**, 6, 609-623.

Valderrama JO, Obaid-Ur-Rehman S and Cisternas LA (1988) Generalized Interaction Parameters in Cubic Equations of State for CO_2 - n-alkane Mixtures. Fluid Phase Equilibria, **40**, 217-233.

Van der Spek A, Thomas A (1998) Neural Net Identification of Flow Regime using Band Spectra of Flow generated Sound. Paper SPE 50640 presented at the European Petroleum conference, The Hague, 20-22 October 1998, proceedings, Vol 2, pp. 93-101.

Van der Waals JD (1873) The equation of state for gases and liquids. Nobel Lectures in Physics, **1**, 254-265.

Vangen G, Carstensen C, Bakken LE (1995) Gullfaks Multiphase Booster Project. Paper OTC 7930 presented at the 27th Offshore Technology Conference, Houston, May 1995.

Vetter G, Wincek M (1993) Performance prediction of twin-screw pumps for two-phase gad/liquid flow. Pumping Machinery ASME, 154, pp. 331-340.

Vetter G, Wirth W, Körner H, Pregler S (2000) Multiphase pumping with twin-screw pumps – understand and model hydrodynamics and hydroabrasive wear. Paper presented at 17th International Pump Users Symposium, March 2000 Houston, proceedings, pp. 153-169.

Vidal J (2003) Thermodynamics: Applications in Chemical Engineering and the Petroleum Industry, Editions Technip.

Vierkandt S (1988) Severe slugging in a pipeline-riser system, experiments and modeling. Ph. D. Dissertation, University of Tulsa, USA.

Villamizar J, Beltran A, Guindi A, Flores J, Afanador L, Aguilar H (2003) Performance Evaluation of a New Gas Handler in an Oil Well with more than 60% Free Gas at the Intake. Presented at the SPE forum on Electrical Submersible Pumps, 2003.

Wallis GB (1969) One dimensional Two-Phase flow. McGraw-Hill, Washington.

Warren P, Al-Dusari K, Zabihi M, Al-Abduljabbar J (2003) Field-Testing A Compact Multiphase Meter – Offshore Saudi Arabia. Paper SPE 81560 presented at the 13th Middle East Oil Show & Conference, Bahrain, 9-12 June 2003.

Wilson GM (1969) A Modified Redlich-Kwong Equation of State, Application to General Physical Data Calculations. Paper **15C**, presented in the AIChE 65th National Meeting, 4-7 May 1969, Cleveland, Ohio.

Winn FW (1957) Physical Properties by Nomogram. Pet. Refiner, **36**, 21, 157-159.

Wong DSH, Orbey H and Sandler SI (1992) An Equation of State Mixing Rule Using Available Activity Coefficient Model Parameters and which allows extrapolation over large ranges of temperature and pressure. Ind. Eng. Chem. Res., 31, 2033-2039.

Wu Z, Xie CG, de Chizelle YK (2000) Microwave Doppler Flowmeter. Patent GB 2,359,435 A.

Wysatta M (1996) Multiphase Pumping: On the Threshold. Journal of Petroleum Technology, February 1996.

Yamazaki Y, Yamaguchi K (1979) Characteristics of cocurrent two-phase downflow in tubes flow pattern, void fraction and pressure drop. J. Nuc. Sci. & Technol. 16, 4, pp. 245-255.

Yocum BT (1973) Offshore riser slug flow avoidance, mathematical model for design and optimization. SPE 4312, European Meeting, London, England.

Zuber N, Findlay JA (1965) Average volumetric concentration in two-phase flow systems. J. Heat Transfer 87, pp. 453-468.

Index

T

U

V

W

Z

Achevé d'imprimer par Corlet Numérique - 14110 Condé-sur-Noireau
N° d'Imprimeur : 69093 - Dépôt légal : avril 2010 - *Imprimé en France*

ASKARI